Structural Repair and Strengthening of Timber-Framed Houses

This book presents structural repairs and strengthening solutions for timber-framed houses with various infills which have been subjected to laboratory tests and translated into easy to understand and apply recommendations. This highlights common mistakes or solutions which are not appropriate to all situations. Historically, work on traditional houses was done by master carpenters, who brought knowledge, techniques, and the access to materials and apprentices. And owners were also involved in the work. But now these trades and skills have almost disappeared. This book serves to set out workable methods in a robust way with scientific validation.

Andreea Dutu is Associate Professor with Habilitation at the Universitatea Tehnică de Construcţii Bucureşti, Romania, and co-author of *Seismic Resistance of Vernacular Timber Frames with Infills: Case Studies from Japan and Romania* (also published by CRC Press).

Mihai Niste is a lecturer at the Universitatea Tehnică de Construcţii Bucureşti, Romania, with research experience in the seismic behavior of traditional timber-framed houses.

Structural Repair and Strengthening of Timber-Framed Houses

Andreea Dutu and Mihai Niste

CRC Press
Taylor & Francis Group
Boca Raton London New York

CRC Press is an imprint of the
Taylor & Francis Group, an **informa** business

First edition published 2027
by CRC Press
2385 NW Executive Center Drive, Suite 320, Boca Raton FL 33431

and by CRC Press
4 Park Square, Milton Park, Abingdon, Oxon, OX14 4RN

CRC Press is an imprint of Taylor & Francis Group, LLC

ISBN: 9781041087786 (hbk)
ISBN: 9781041088509 (pbk)
ISBN: 9781003647249 (ebk)

DOI: 10.1201/9781003647249

Typeset in Times
by Newgen Publishing UK

Contents

Preface

When the first idea of the book appeared, it was coming as a solution to the lack of engineering studies on the strengthening techniques for traditional timber frames with infills.

There are several good books, made by architects, where recipes and local solutions are given in detail, but they lack some engineering factors, such as increase in strength, quantities influence in the materials recipes, etc.

The book is based on the StrongPa project, implemented at the Technical University of Civil Engineering Bucharest between 2022 and 2024, where we wanted to scientifically validate several strengthening solutions for traditional timber frames with various infills. As we are structural engineers, we focused on our specialty and wanted to see which method is more effective in terms of strength and stiffness. And we also have considered the easiness in their application and finding the materials, and of course, cost. Another factor we had considered was whether the materials were natural or not, thus if compatible with the traditional houses' existing materials. In our team, we also had architects, militating for the compatibility between materials, but since our main specialty was structural engineering, this had priority in our studies.

Thus, in the solutions provided, besides compatible solutions we felt obliged to consider also the most common ones applied in real life, many of them involving cement mortar, which is known to produce compatibility issues long term for the houses, mainly due to the different physical and mechanical properties. Cement mortar is rigid and brittle, while timber and earth infill are flexible and deform with moisture and temperature changes. When the timber swells or shrinks (as it naturally does), the rigid cement cannot accommodate movement and thus cracks form, breaking the bond and allowing moisture to ingress. Cement is impermeable or has low vapor permeability, while earth and timber need to breathe. Cement mortar traps moisture at the timber interface, leading to timber rot and fungal decay, disintegration of the earthen infill over time or salt crystallization, and internal stresses. Cement has different thermal expansion characteristics than both timber and earth, so changes in temperature or humidity cause differential movement, worsening cracks and separation. Cement is hard to remove and often damages the adjacent historic fabric (timber and earth) during repair. Traditional lime or clay mortars, by contrast, are reversible and compatible with natural materials.

Nevertheless, in our proposed solutions, we have also successfully validated versions which involve only timber, which is perfectly compatible with the existing materials.

In the case study, we aimed to provide an engineering perspective on the chronological steps of an intervention on a traditional timber frame. We gave a lot of details and explained the technical terms so that even if the owner decides to do it by him/herself, the correct methods should be known and well understood, so we can avoid costly mistakes in the future, both from the point of view of seismic resistance (Romania is a seismic prone country), and from the durability point of view.

As mentioned in Chapter 5, although the steps described involve cement mortar use, depending on the owner and architect/engineer that implement such strengthening works, they can choose to substitute it with another strengthening solution, as presented in Chapter 4, suitable for a specific house, location, and owner, but keeping the main intervention steps the same.

Acknowledgments

This book is the outcome of my latest research project, which I implemented in UTCB and for which, with the great help of Prof. Shoichi Kishiki from Institute of Science Tokyo, I had the opportunity to work with two of his brightest doctoral students, Dr. Ichiro Hirano and Dr. Sometrey Mey. Although the target of the project was to make experiments on 9 walls, due to the quality and working rhythm of Ichiro and Sometrey, we could test 17 walls in just two months. It was a wonderful experience, which although involved tough work, I could see a dream come true: working in Japanese style and with similar quality in a Romanian university! I thank them all for the help and support along with the StrongPa project.

I would like to thank my family for their patience and support. My husband, my mother, and my late father, they all encouraged me to go ahead and pursue my dreams no matter what. My three sons, who inspired me and empowered me.

I am thankful to all those from whom I have learned in my career, and especially to those who believed in me and never stopped doing so. I acknowledge all the opportunities my mentors offered me and their kind collaboration despite the distance. It all started with Prof. Joao Gomes Ferreira who kindly welcomed me into his team and thus I started studying the topic of timber frames with various infills. He taught me how to write my first paper and showed me how a functional education system can weigh tremendously in a young researcher's life. But the most valuable thing he taught me was that supervisors can be friends with the students, even though their competence is way beyond the students'.

Prof. Hiroyasu Sakata offered me the opportunity to go to Japan and be near greatness (in his lab), from which I learned and with which I lived, and thus my values and standards forever changed. It still is an honor to present myself as an ex-postdoc of Sakata laboratory. After my return to Romania, Prof. Akira Wada never stopped teaching me greatness and many times lit my path that had become dark.

Special thanks to my co-author and friend, Prof. Mihai Niste, who helped and worked with me even though sometimes things got complicated. And also, to my international friends (too many to name them all here), who keep me anchored into the real research world.

The StrongPa project is financially supported by the Romanian National Authority for Scientific Research and Innovation, CNCS – UEFISCDI, under project number PN-III-P2-2.1-PED-2021-1428 – Experimental Demonstrative Project (UEFISCDI Romania), 2022–2024.

I also acknowledge the great support of MSL funding of Tokyo Institute of Technology (now Institute of Science Tokyo), supporting the research activity with many laboratory items and experimental activities.

About the Authors

Andreea Dutu (after marriage CASUTA) is a structural engineer, who graduated with bachelors, masters and PhD degrees from the Technical University of Civil Engineering in Bucharest, Romania. Her main research background is in the seismic evaluation of traditional timber buildings with various types of infills (masonry, clay, logs, etc.). She also has a passion for seismic isolation, energy dissipation, and vibration control of structures and she was elected as a member of the Board of Directors in the ASSISI Society (Anti-seismic Systems International Society).

Starting from her PhD studies, she has been enrolled in several research stages around the world (Portugal, France, Turkey, China), but the longest one was in Japan, at the Tokyo Institute of Technology, where she stayed for two and a half years as a post-doctoral researcher. In Japan, under the supervision of Prof. Hiroyasu Sakata and Prof. Yoshihiro Yamazaki, her way of thinking profoundly changed from both a professional and personal point of view.

When she returned to Romania, she had a dream to apply what she had learned in Japan, meaning to never give up in pursuing her objectives (to do research as in Japan) and to embrace trial-and-error and accept it as a way of learning and growing. So far, she obtained and successfully implemented three research grants in the home university where she also teaches several courses and could study the traditional houses from the engineering point of view for the first time in Romania, conducting an extensive experimental program on sub-assemblies (walls, connections, component materials, etc.).

She is an associate professor and obtained her Habilitation in 2025 from the Technical University of Civil Engineering in Bucharest, Romania.

Mihai Niste is a lecturer at Technical University of Civil Engineering in Bucharest, Romania, and also a fire fighting officer in the reserve with extensive research experience in the seismic behavior of traditional timber-framed houses. Mihai graduated from the same university, Faculty of Hydrotechnics the Fire Officers School, the Master's program in Special Technologies in Construction, and the Faculty of Civil Engineering. He previously worked in the Military Firefighters Corps in structural design and then at the formwork and scaffolding company PERI GmbH. He now teaches construction engineering (technology) and also has great experience on the construction site.

1 Introduction

Traditional timber houses in Romania were scarcely studied from the engineering point of view, even though, after the strongest seismic events that hit the country in 1977 and 1986, the reports showed that they were rarely damaged seriously, which was an obvious point of interest. One possible explanation for this lack of strong damage was found in the frequency content of the local ground motions recorded during these earthquakes, which was not of the nature of affecting small one-story residential houses, sensitive to short-period components. However, given the complex seismicity of Romania, with the Vrancea subcrustal source that generates 2–3 M>7 earthquakes per century and with several other crustal sources generating smaller magnitude events, a general conclusion is difficult to formulate.

Traditional residential houses represent valuable assets from the heritage point of view and an important part of the cultural identity. In Romania, according to the building census in 2021, 28% of the built environment (residential) consists of traditional houses, with 6% timber structures and 17% timber frames with infills along with adobe houses. Most of these date from the period 1946–1960, while others are historical buildings in rural areas, which resisted time, being maintained and inhabited by successive generations. Thus, this type of buildings represents a quite large proportion of the total building stock. Today, many people are aware of the need of their strengthening and preservation; however, they lack systematic knowledge on the specific methods and practices required.

In other countries, such as Portugal [1], in which similar vernacular construction techniques are applied, several methods were proposed for strengthening timber frames with masonry infills. Still, even if the construction principles of the structural types similar to "*paianta*" are the same, due to the variation of detailing, specific to a country and sometimes only to a limited ethnographic area, the strengthening methods from one country cannot be always applied to another.

The StrongPa project, funded by the Romanian National Authority for Scientific Research and Innovation, CNCS-UEFISCDI, under project number PN-III-P2-2.1-PED-2021-1428, aims at providing scientifically validated solutions for the strengthening of timber-framed traditional houses, infilled with various types of materials.

DOI: 10.1201/9781003647249-1

Many strengthening solutions were considered, some of which were already investigated by other researchers, although applied on different structural types, such as: used car type straps [2–3], disperse fiber reinforced mortars [4–7], near surface mounting rebars [8], polymeric meshes [6–10], and low-cost seismic isolation [11–13].

However, the decision on which to test for the *paianta* was made based on the feasibility of the solutions to be applied in Romania by owners of *paianta* houses. Thus, the ones considered more suitable for the Romanian culture, experience, and current requirements were selected to be experimentally investigated.

2 Types of Timber-Framed Houses

The timber frames with infills are of many types, even within the same country. An example is Romania, where we have done the field investigation and found five types of timber frames with different infill types, the types being in fact differentiated by the infill material. In Figures 1–5 are shown the infill types found. In Figure 6 the possible carpentry connections that may be found in real houses for the timber frame. The main structural components of such a house are shown in Figure 7, notions that we will mention later in the book. More details on the original execution of such infills are given in [14].

DOI: 10.1201/9781003647249-2

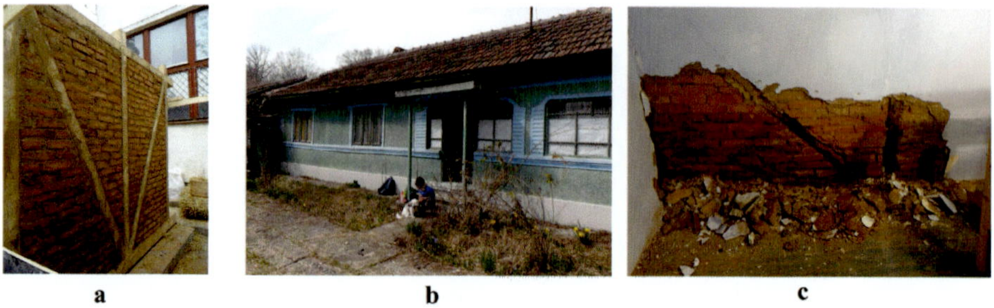

FIGURE 1 *Paianta* house type 1 (timber frame with clay brick masonry infills): (a) wall specimen built in the laboratory; (b) the house taken as a reference for the wall reproduced in the laboratory; (c) the structural system of the house.

FIGURE 2 *Paianta* house type 2 (timber frame with wattle and daub infills): (a) wall specimen built in the laboratory where the wattles are interlaced and laid horizontally; (b) wall specimen built in the laboratory where the clay mortar is applied on the wattles; (c) the real house taken as a reference for the wall specimens reproduced in the laboratory; (d) real house's wall with the wattles laid vertically; (e) the whole house with wattles laid vertically and clay mortar lacking on some areas.

FIGURE 3 *Paianta* house type 3 (timber frame with horizontal timber laths and mud and straw infills): (a) wall specimen built in the laboratory with horizontal laths made of timber battens, not branches; (b) adding the clay mortar within the laths is done by throwing chunks of mixed clay and straws inside the space created by the laths on both sides of the wall; (c) the real house taken as a reference; (d) the whole reference house.

FIGURE 4 *Paianta* house type 4 (timber frame with horizontal timber logs infills): (a) wall specimen built in the laboratory with horizontal logs infill; (b) the real house taken as a reference; (c) another example of such a house.

FIGURE 5 *Paianta* house type 5 (timber frame with AAC brick masonry infills): (a) wall specimen built in the laboratory with AAC bricks infill; (b) a real house taken as a reference; (c) another example of such a house.

FIGURE 6 Types of traditional carpentry joints in vernacular timber-framed houses: (a) mortise and tenon joint; (b) cross-halved joint; (c) stepped joint; (d) dove-tail joint.

FIGURE 7 *Paianta* house component elements.

3 Degradations and Their Causes

Most of the degradation in this type of houses comes from poor maintenance, thus not so much from earthquake damage. These are generally not threatening to the structural stability of the houses and can rather be repaired easily. On the other hand, in earthquake prone areas, the types of earthquake damage should be considered, and treated, to prevent the collapse of the house or the functionality disruption for long periods after a major earthquake. The degradations, depending on their source, can be classified as follows:

1. Natural (water, soil, weather, biological).
2. Structural (foundations, joints, materials, loads).
3. Human (poor design, poor maintenance, bad interventions).
4. Accidental (earthquake, fire, floods, storms).

Further, possible degradations are pointed out very briefly, as could be found in the field investigation and experience of the authors. A detailed list of possible degradations can be found in [14], [15].

3.1 DEGRADATIONS UNDER NORMAL USE (NO EARTHQUAKE)

- Moisture/water infiltration
    - Causes: rain, snow melt, flooding, capillary rise.
    - Effects: timber rot (especially mudsills), fungal attack, insect infestation, plaster peeling, biological decay.
    - → *Note*: drainage, chemical treatment, replacing degraded timber.
- Foundation problems
    - Lack of foundation/weak foundations (mud mortar, dry stone).
    - Differential settlements → cracks in walls, inclinations, partial collapse if abandoned.
    - → *Note*: stabilize soil, improve/replace foundation.
- Structural conformation errors
    - Weak/gapped connections.

DOI: 10.1201/9781003647249-3

  - Asymmetrical or poorly connected diagonal braces → torsion.
  - Infill sensitive to out-of-plane failure, plaster detachment.
  - → *Note*: reinforce connections, bracing, meshes.

3.2 DEGRADATIONS DUE TO EARTHQUAKES

- Out-of-plane failure of infills (Haiti, Greece, Japan).
  - Infills fall outside, dangerous for people outside.
- Cracking and plaster loss at corners, frame-infill contact.
- Roof issues in Japan: heavy ceramic roofs cause leaning/collapse in subsequent quakes.
- Connection failures: mudsill–post detachments → risk of collapse.
  - → *Note*: strengthen infill-frame connection, improve joints, reduce roof weight.

3.3 ACCIDENTAL DEGRADATIONS

- Landslides/soil creep → settlements, inclined cracks, out-of-plane deformations.
- Floods → accelerated biological decay, plinth masonry degradation.
- Storms → roof cover/rainwater system detachment.
- Fire → partial/total destruction of timber elements.

3.4 ENVIRONMENTALLY INDUCED DEGRADATIONS

- Humidity & freeze-thaw → masonry spalling, plaster peeling, cracks, timber instability.
- Biological growth (mold, algae, lichens) → decay of timber and mudsills.
- Rodents/birds → reduce timber section, bacterial attack.

3.5 MECHANICALLY INDUCED DEGRADATIONS

- Poor/collapsible soil → differential settlements, humidity rise.
- Undrained soil → capillary action → timber rot.

3.6 IMPROPER CONFORMATION DEGRADATIONS

- No/weak foundations → cracks, timber detachment, settlements.
- Overloading from changes in use/heavy finishes → cracks, bending.
- Weak joints or defective timber → crooked, cracked, shrinking, disintegration (especially softwood).
- Short eaves/lack of pavement → humidity exposure → moss, algae, decay.
- North-facing walls → high humidity → decay.

3.7 POOR MAINTENANCE DEGRADATIONS

- Neglected biological attacks → progressive decay.
- Vegetation and tree roots → moisture, foundation damage.
- Abandonment/lack of heating → high indoor humidity.
- Clogged or missing drainage → plaster, infill, timber degradation.
- Missing windows/doors → pests, humidity, vandalism.

3.8 INAPPROPRIATE INTERVENTIONS DEGRADATIONS

- Concrete additions (plinth, plaster, walkways) → capillary rise, dampness, biological decay.
- Bad joint strengthening/static scheme alterations → detachments, cracks, bent elements.
- Incompatible materials (cement, asphalt board, vapor-proof paints, plastic thermal insulation) → moisture trapping, timber decay.
- Heavy traffic vibrations nearby → wall cracking, settlements.

4 Strengthening Solutions for Timber-Framed Houses

The strengthening solutions proposed and tested within the StrongPa project [16] are further described, initially all of them including the ones that turned out to be less effective, and later, for the application guiding, for the most successful ones details will be given on how to actually apply them, as we have experienced them in the experimental lab. The objectives of the project included, besides the mechanical characterization of each tested specimen, testing the execution procedure adopted and finding the best and easiest solutions to apply the proposed strengthening solutions, considering the local access to the materials.

Two types of solutions were tested, as we will call them further "dry" and "wet", the later ones involving the application of materials that needed a certain period to dry (i.e.: one month for those with cement, and at least two months for those with clay). The "dry" solutions aimed to use readymade materials, easy to get from the Romanian (thus European) market, and also easy to apply. The "wet" solutions even if involve more time to cure, aimed similar objectives, to be easily applied even in a "do-it-yourself (DIY)" regime, which is preferred in the rural areas, at least in Romania. Want it or not, this is the reality to which we have to adapt, and if we, specialists, want to prevent execution mistakes that can be life-threatening in an earthquake, we have to provide the general population correct information, which is also scientifically proved, that can in many cases be implemented by the owners themselves. Of course, we recommend contracting an architect–engineering team, as the results are much better when done professionally. But in the (many) cases when in rural area people choose to make improvements to their houses, due to budget limitations, they tend to skip the professional contracting part. This comes with major disadvantages in many cases, as for such work (DIY) there is no warranty when something goes wrong, and the owners cannot recover their original investment (even if small).

Through field investigations and discussions with other specialists (architects and engineers), we have noticed that the most common way to strengthen a traditional house is to apply on the walls, on both sides, a cement mortar reinforced with welded wire meshes or some mesh used for plastering, which resembles to reinforced concrete jacketing, sometimes reaching 2 cm thick on one side. Architects complain that this solution is not compatible with the existing materials (clay/earth) and in time it produced an advanced biological decay, especially at

DOI: 10.1201/9781003647249-4

the interface between the two materials, which also have adherence problems. This, in the end, causes disintegration of the original natural materials of the walls and an unhealthy environment indoors.

Since we are engineers, and our main target was structural behavior under lateral loading, we focused only on efficiency in terms of strength, stiffness, and deformation without collapse for the proposed solutions. The compatibility between materials has not been considered yet, as our equipment is not able to do experiments on this topic yet.

Although a previous project, TFMRO [17], developed the first experimental program on *paianta* houses, the variability of the test results still represented a concern, especially for walls involving the use of clay/earth. For this reason, more specimens needed to be tested to understand how the properties vary, depending on the workmanship or on some small differences in the construction details. So, within the next project, implemented at the Technical University of Civil Engineering Bucharest (UTCB), 15 walls were tested in a cyclic static regime on the reaction frame donated by the Japan International Cooperation Agency (JICA), and located in the structural testing laboratory of UTCB.

All the five types of *paianta* houses, described earlier in Chapter 2, were investigated experimentally and, for the first three types, which according to the field investigation are the most largely used [18], three identical walls were built, in an attempt to assess the variability given by the material properties and workmanship differences. Two identical wall specimens were constructed and tested for Type 4, while only one specimen was constructed and tested for Type 5.

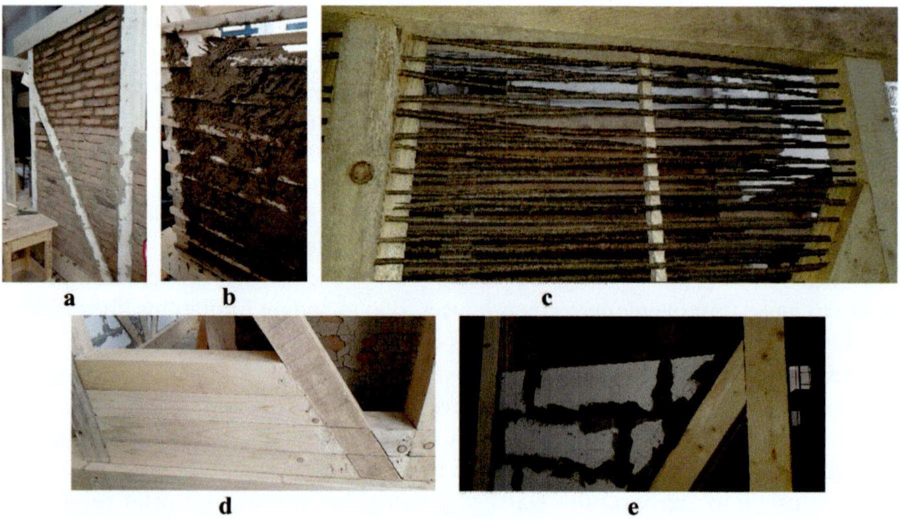

FIGURE 8 The types of infills within the StrongPa project: (a) fired earth brick infill with earth mortar; (b) horizontal timber strips and earth mixed with straws infill; (c) wattle and daub infill on which earth mixed with stray will be applied; (d) timber logs infill; (e) AAC masonry infill with cement mortar.

After the execution of the walls, they were first tested applying the EN12512 loading protocol on the walls without any strengthening solution applied. After this test, the strengthening was applied and then the same loading protocol was applied on the strengthened wall.

The reference timber frame, called further on "the main timber frame", had the dimensions and characteristics as following: the cross section of posts, beams and bracings was 120x120 mm; the connections between them were nailed (plain); total height was 2400 mm (with bottom beam and top plate of 200 mm height); and length was 3000 mm (with 2340 mm between axis of marginal posts) [16].

Figure 8 shows the types of infills during the construction of the walls for a better understanding of the materials and details.

4.1 DRY STRENGTHENING SOLUTIONS

In Figure 9 the materials used for strengthening and the positions where they were applied within the tested walls are presented.

Table 1 shows the materials used for each strengthening solution, the estimated cost and the envelope comparison on the positive loading on the wall without and with strengthening. It may be noted that the strengthening was always done after the wall was tested in the non-strengthened state. On the graphs, the red line is the envelope of the strengthened wall, while the black line is that of the non-strengthened one. The types of walls on which the strengthening solutions were applied are in accordance with [19], and in this way, the specimen's name and type of infill can easily be identified.

4.2 WET STRENGTHENING SOLUTIONS

In Figure 10 the "wet" strengthening solutions are briefly presented, and later in the book details will be given on the full execution process.

The test results are shown in Table 2, together with the materials involved and their cost, including the workmanship. The curve (envelope) comparison is also shown, the red line is the strengthened wall, while the black line is the non-strengthened wall.

4.3 COST ANALYSIS FOR STRENGTHENING SOLUTIONS APPLIED TO A WALL WITH DIMENSIONS OF 2X2 M

For each strengthening solution, the total price was calculated, which includes the price of materials and the related workmanship (labor). The labor cost was considered the market practice, according to which the price of labor is at a similar level to that of materials. For the analysis in Table 3, this price was considered depending on the skill level required for each solution. For levels 1 and 2, the labor cost is 80% of the total cost of materials. For levels 4 and 5, the labor cost is 120% of the total cost of materials. For level 3, the labor price is equal to that of materials.

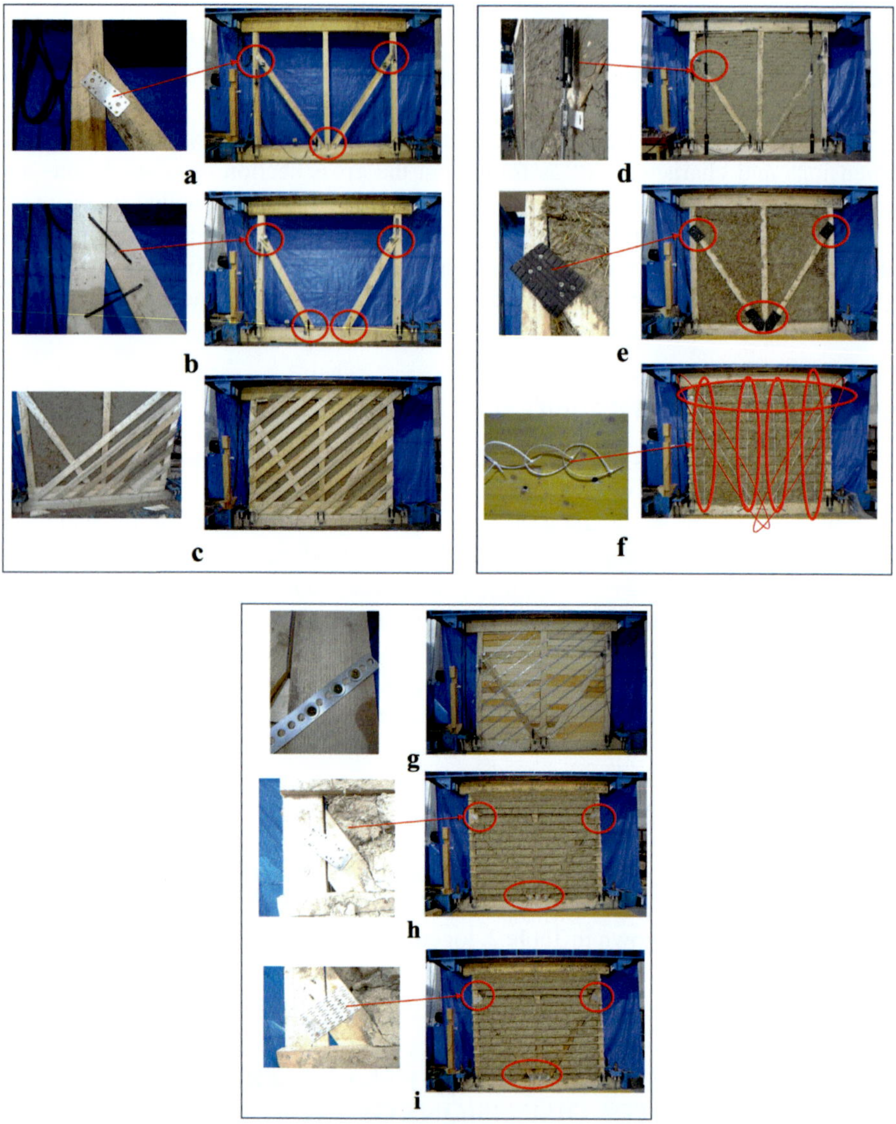

FIGURE 9 Materials used for strengthening (dry) and their position on the wall: (a) metal plates and corner plates; (b) clamps (profiled or not); (c) diagonal planks (TRAROM); (d) steel tie rods, having a pretensioning mechanism with turnbuckles; (e) used tire straps; (f) cable tie bands chains on diagonals, vertical, and horizontal; (g) fixing band applied obliquely and orthogonally on both sides; (h) metal plates and corner plates; (i) punched metal plates and corner plates.

TABLE 1
Details of the applied dry strengthening solutions and their efficiency

No.	Solution	Specimen and type of infill	Materials applied	Cost [RON]	Envelope comparison
1	Metal plates	S2 (no infill)	Flat metal plate 55x135x2 mm Corner metal plate 70x70x55x2 mm Screws 5x70 mm	141	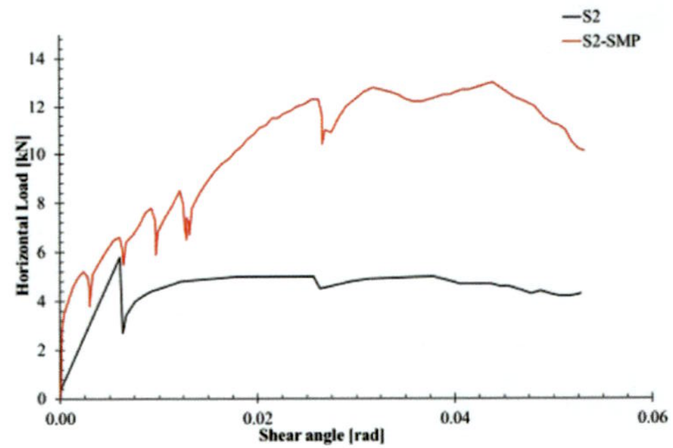
2	Clamps	S3 (no infill)	Smooth clamps φ6 of 200 mm Profiled clamps (PC60) φ8 of 200 mm	58	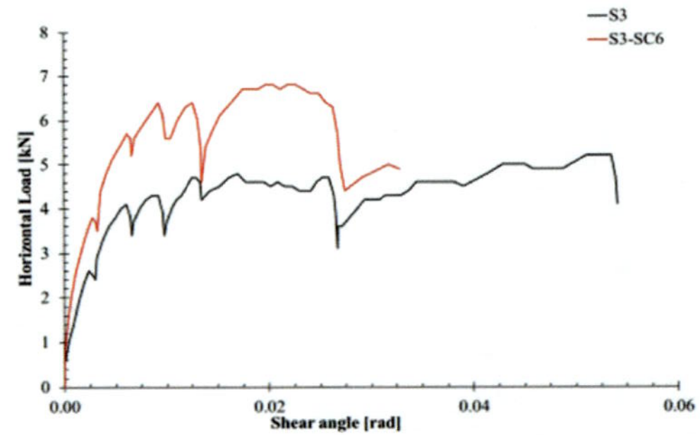

(*continued*)

TABLE 1 (Continued)
Details of the applied dry strengthening solutions and their efficiency

No.	Solution	Specimen and type of infill	Materials applied	Cost [RON]	Envelope comparison
3	TRAROM diagonal planks	S4-2 (clay brick masonry infill)	Timber planks 120x20x4000 Screws 6x80 mm	806	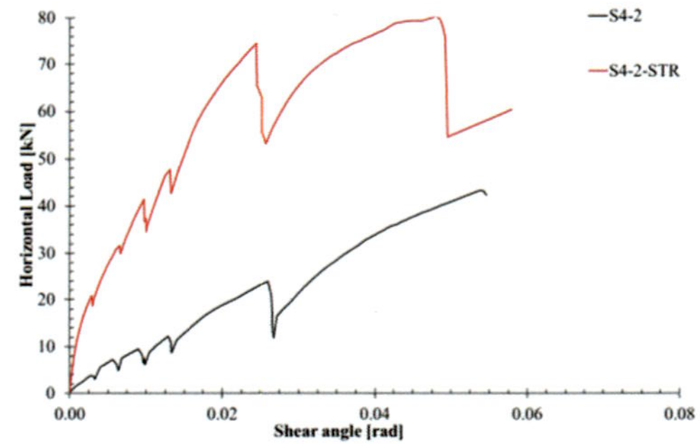
4	Steel tie rods	S4-1 (clay brick masonry infill)	Metal plate 5x60x2000 Big nuts 12x36 Small bolts M10x25 Small nuts M10 Long bolt 12 mmx1000 Flex disks Drill for metal 12 mm Bolts M10x50 mm Flat washers M10 Nut M10 Flat bar Turnbuckles	1075	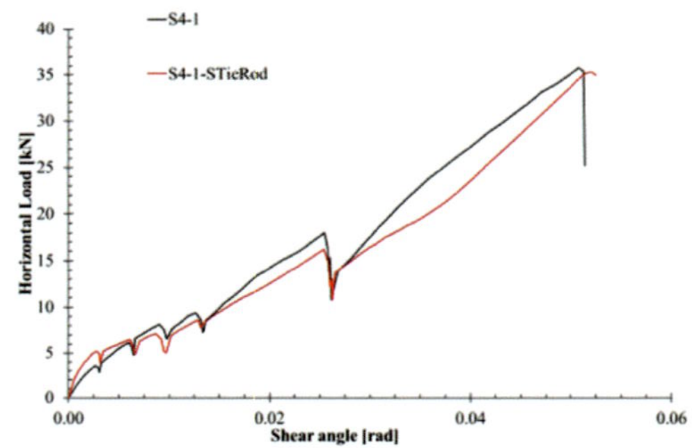

5	Used tyre straps	S6-1 (wattle and daub infill)	Washers Recycled used tyre Screws 6x80 mm	49	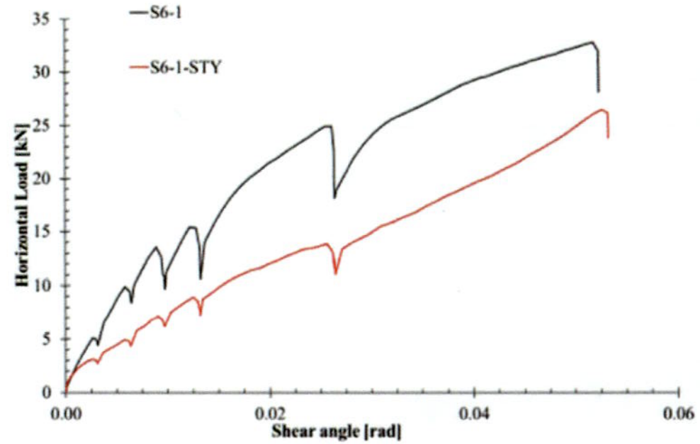
6	Cable tie bands chains on diagonals and vertical	S5-1 (horizontal laths and mud mixed with straws infill)	Cable ties 500x8 mm, 50 pieces/pack Cable ties 550x9 mm, 50 pieces/pack Screws 6x80 mm	278	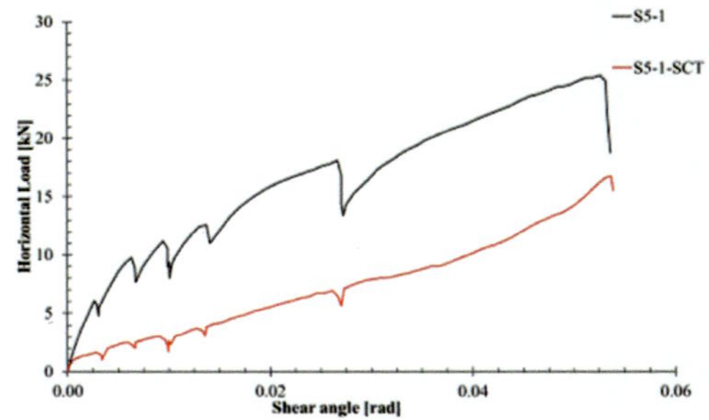
7	Fixing band	S9-2 (horizontal logs infill)	fixing band 12m, 17mm, Lidl Fixing band 10m, 25 mm, Dedeman Washers diameter 6 mm Screws 6x80 mm	660	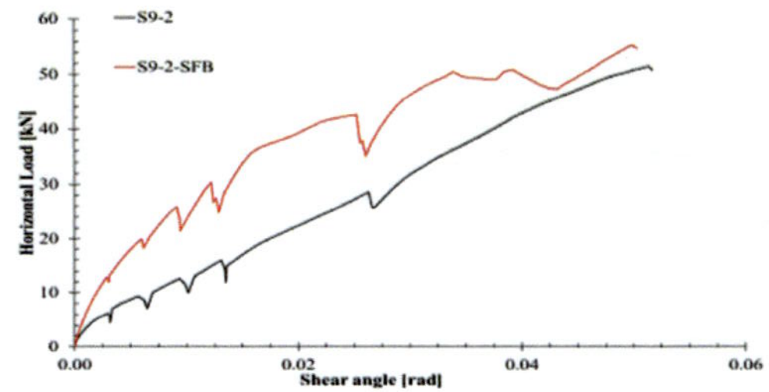

TABLE 1 (Continued)
Details of the applied dry strengthening solutions and their efficiency

No.	Solution	Specimen and type of infill	Materials applied	Cost [RON]	Envelope comparison
8	Metal plates	S5-3 (horizontal laths and mud mixed with straws infill)	Flat metal plate 55x135x2 mm Corner metal plate 70x70x55x2 mm Screws 5x100 mm	141	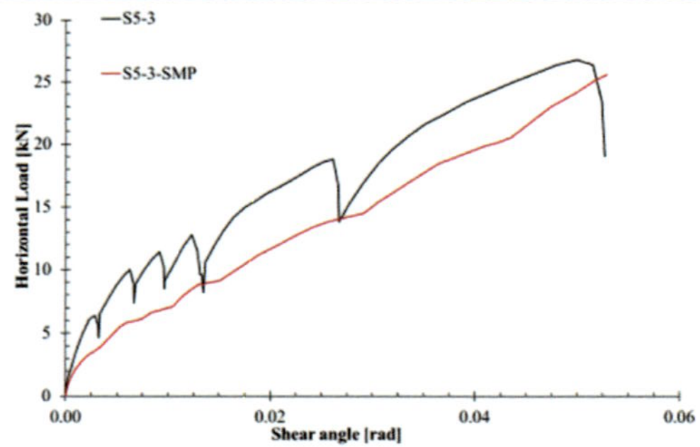
9	Punched metal plates	S5-3 (horizontal laths and mud mixed with straws infill)	Punched metal plates 77x150 mm Punched metal plates 102x200 mm Screws 5x100 mm	120	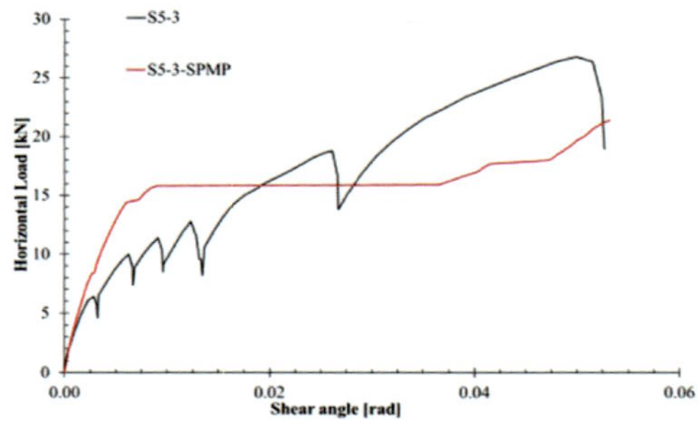

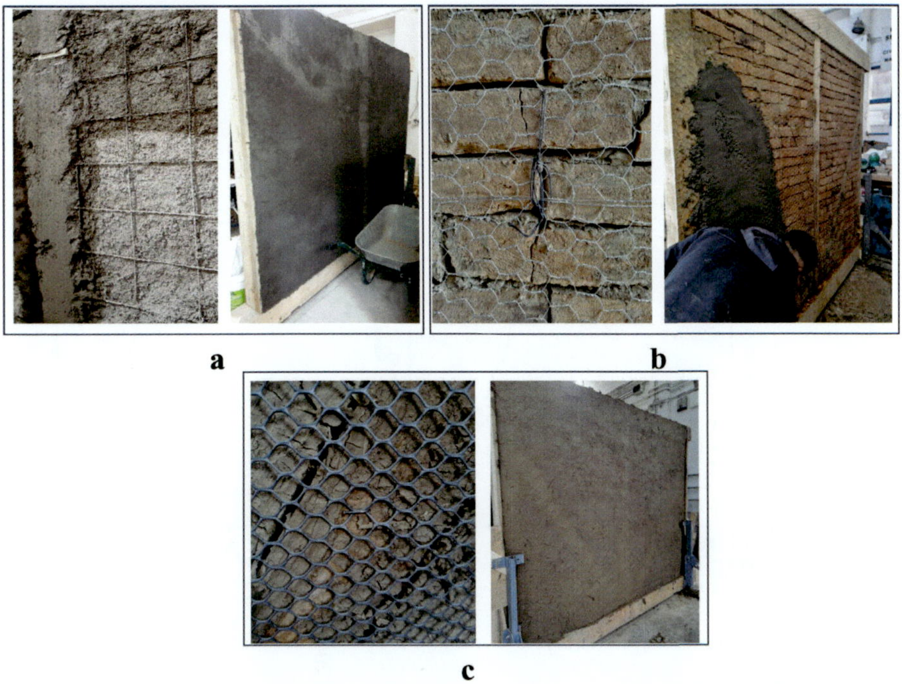

FIGURE 10 Materials (wet) used for strengthening (left) and their application on the wall: (a) steel welded wire mesh and cement mortar; (b) rabbit wire mesh and cement mortar; (c) plastic mesh and clay/earth mortar.

4.4 PROPOSALS FOR STRENGTHENING SOLUTIONS FOR TIMBER FRAMES WITH DIFFERENT TYPES OF INFILLS

4.4.1 Dry Solution TRAROM—Using Unplanned Wooden Planks Applied Obliquely to the Wall

The solution involves first applying a row of adjustment planks to overcome the irregularities/protrusions of the infill, so that there is no need for hammering the entire surface of the infill (to increase adherence with the newly added material), which would take a lot of time and generate a significant amount of dust. For a small number of protrusions of the infill, a hatchet is used to remove those parts (Figure 11d). A 20 mm thick plank is sufficient to be applied along the contour of the timber frame of the wall, serving as a support layer for attaching the diagonal planks (also having a thickness of 22 mm). The areas where the adjustment/support plank pieces are applied are marked in red in (Figure 11e), so that the main planks can be fastened to the adjustment row using timber screws. The planks were un-processed and un-treated, made of local fir, thus cheap and easily available on the local market. The screw nails were for timber structures, ø6 x 80mm, zinc coated. They were inserted using a cabled drilling machine. Generally, cabled drilling machine prevents stops due to low battery, especially when the drilling machine is not the best on the market.

TABLE 2
Details of the applied wet strengthening solutions and their efficiency

No.	Solution	Specimen and type of infill	Materials applied	Cost [RON]	Envelope comparison
1	Welded wire mesh and cement mortar	S8 (AAC brick masonry infill)	cement [bag], welded mesh 2000x5000 mm [sheet], sand [m3], steel wire 1.18 mm [m], 1.8x50 mm nails [kg]	1650	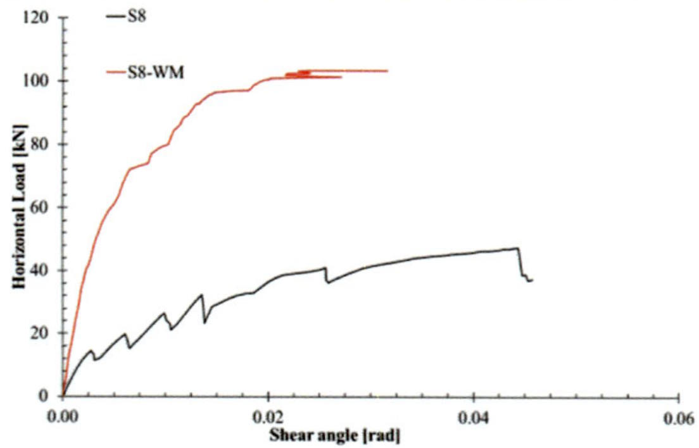
2	Rabbit wire mesh for plaster and cement mortar	S4-2 (clay brick masonry infill)	Rabbit wire mesh 1x25 m roll, 0.8 mm diameter, cement [bag], sand [m3], 1.8x50 mm nails [kg], steel wire 1.18 mm [m]	1704	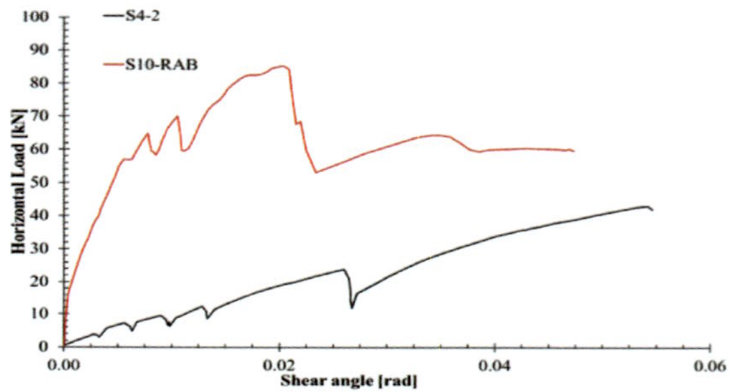

3	Plastic mesh and clay/earth mortar	S6-2 (wattle and daub infill)	Plastic mesh 1x30 m, 1.8x50 mm nails [kg], steel wire 1.18 mm [m], sand [m3], straws	710	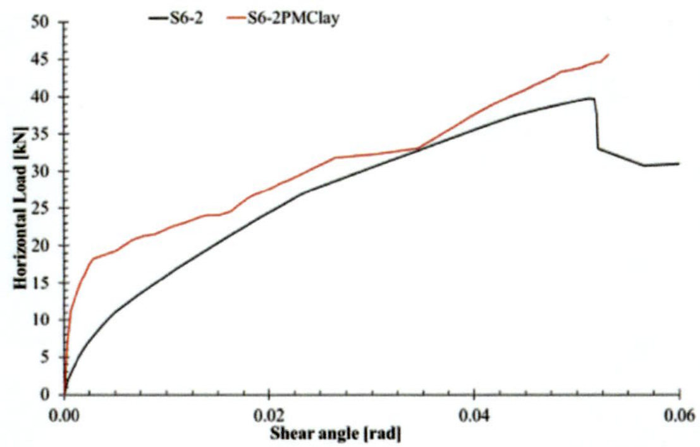

TABLE 3

Costs of materials and workmanship per each strengthening solution considered in the StrongPa project

No.	Solution / specimen name	Pieces type	Total price/ material	Total price/ solution [RON]	Hours of labor [h]	No. people involved	Level of skills (1-low skill to 5-high skill)	Price of labor [RON]	Total price per solution/ wall [RON]
1	**Metal plates / S2-SMP**	Flat metal plate 55x135x2 mm	16	78.2	2	2	1	63	141
		Corner metal plate 70x70x55x2 mm	44						
		Screws 5x70 mm	18						
2	**Clamps / S3-SC6/ S3-SC8**	Smooth clamps φ6 of 200 mm	14	32.4	0.5	2	1	26	58
		Profiled clamps (PC60) φ8 of 200 mm	18		0.5	2			
3	**TRAROM diagonal planks / S4-b-STRAROM**	Timber planks 120x20x4000	321	447.6	8	1	2	358	806
		Screws 6x80 mm	126						
4	**Steel tie rods / S4-a-TieRods**	Metal plate 5x60x2000	78	489	10	1	5	586	1075
		big nuts 12x36	32						
		small bolts M10x25	16						
		small nuts M10	5						
		Long bolt 12 mmx1000	112						
		flex disks	28						
		drill for metal 12 mm	53						
		bolts M10x50 mm	9						
		flat washers M10	2						
		nut M10	2						
		flat bar	78						
		turnbuckles	73						

5	**Used tyre straps / S6-a-STY**	Washers	7	27	3	1	1	22	49
		Recycled used tyre	0						
		Screws 6x80 mm	20						
6	**Cable tie bands chains on diagonals and vertical/ S5-a-SCT**	Cable ties 500x8 mm, 50 pieces/pack	74	154	2	2	1	123	278
		Cable ties 550x9 mm, 50 pieces/pack	71						
		Screws 6x80 mm	8						
7	**Fixing band / S9-b-SFB**	fixing band 12m, 17mm, Lidl	60	413	6	2	1	248	660
		Fixing band 10m, 25 mm, Dedeman	132						
		Washers diam 6 mm	45						
		Screws 6x80 mm	176						
8	**Metal plates / S5-c-SMP**	Flat metal plate 55x135x2 mm	16	78.2	2	2	1	63	141
		Corner metal plate 70x70x55x2 mm	44						
		Screws 5x100 mm	18						
9	**Punched metal plates / S5-c-SPMP**	punched metal plates 77x150 mm	16	60	1	2	3	60	120
		punched metal plates 102x200 mm	26						
		Screws 5x100 mm	18						
10	**Welded wire mesh and cement mortar /S8-PS**	cement [bag]	140	825	8	2	3	825	1650
		welded mesh 2000x5000 mm [sheet]	45						
		sand [m3]	600						
		steel wire 1.18 mm [m]	29						
		1.8x50 mm nails [kg]	11.22						

(continued)

TABLE 3 (Continued)
Costs of materials and workmanship per each strengthening solution considered in the StrongPa project

No.	Solution / specimen name	Pieces type	Total price/ material	Total price/ solution [RON]	Hours of labor [h]	No. people involved	Level of skills (1-low skill to 5-high skill)	Price of labor [RON]	Total price per solution/ wall [RON]
11	**Rabbit wire mesh for plaster and cement mortar / S10**	Rabbit wire mesh 1x25 m roll, 0.8 mm diameter	72	852	8	2	3	852	1704
		cement [bag]	140						
		sand [m3]	600						
		1.8x50 mm nails [kg]	11.22						
		steel wire 1.18 mm [m]	29						
12	**Plastic mesh and clay/earth mortar / S6-b-PMClay**	Plastic protection mesh 1x30 m	132	355	8	2	3	355	710
		1.8x50 mm nails [kg]	11.22						
		steel wire 1.18 mm [m]	29						
		sand [m3]	100						
		straws	70						

FIGURE 11 Application of the TRAROM strengthening solution with diagonal planks: (a) The flatness of the wall is checked, particularly in the infill area; (b) it is tested whether the thickness of the plank pieces, in our case 22 mm, is sufficient to space the row of applied planks far enough not to be obstructed by any irregularities in the infill; (c) the wall before applying the adjustment/support layer; (d) a hatchet is used for minor trimming/adjustments of the infills' surface level; (e) the position of the plank pieces for adjusting the irregularities in the filling; (f) after completing the row of adjustment planks, the diagonal planks are applied, starting with the one on the diagonal (the longest one); g) final layout of the TRAROM solution.

Although not tested yet in the out-of-plane direction, this strengthening solution is also improving the out-of-plane deformation of the infills (a frequent damage in earthquakes for such vernacular houses with timber frames and infills), keeping them inside the plank structure.

For one side of the wall, we used about ten pieces of planks, having the dimensions of 120 (width) x20 (thickness) x4000 (length), in mm. The number of screws was around 200 pieces (one box).

4.4.2 Dry Solution with Fixing Band

The solution involves applying fastening bands diagonally across the wall. The fastenings are made using timber screws, and it is recommended to use at least two electric screwdrivers for the screwing operation. These screwdrivers should be corded, not battery-powered, to avoid frequent interruptions during work (waiting for batteries to recharge). The solution is particularly suitable for *paianta* houses with timber logs infills, where fastenings must be made on each horizontal log of the infill.

On the main timber frame (posts, bottom beam, upper beam, and diagonals), it is recommended to make the screw fastenings at a minimum of two points, preferably three. Screws should be used with washers to avoid inaccurate fastening in the holes of the fixing bands. Fastening to the bottom beam (in our setup representing the mudsill) involved bending the band from the wall onto the beam.

For greater strength, the band should be as wide as possible. In the laboratory, the solution used two band widths: the widest (25 mm) was applied on the central diagonal, while the narrower one (17 mm) was used in other areas, which experience less load in the case of a wall subjected to lateral forces. In order to cut the band to the desired length we used special scissors (Figure 12a) for cutting metal sheet, which is strong enough to cut the band and does not require too much effort from the user.

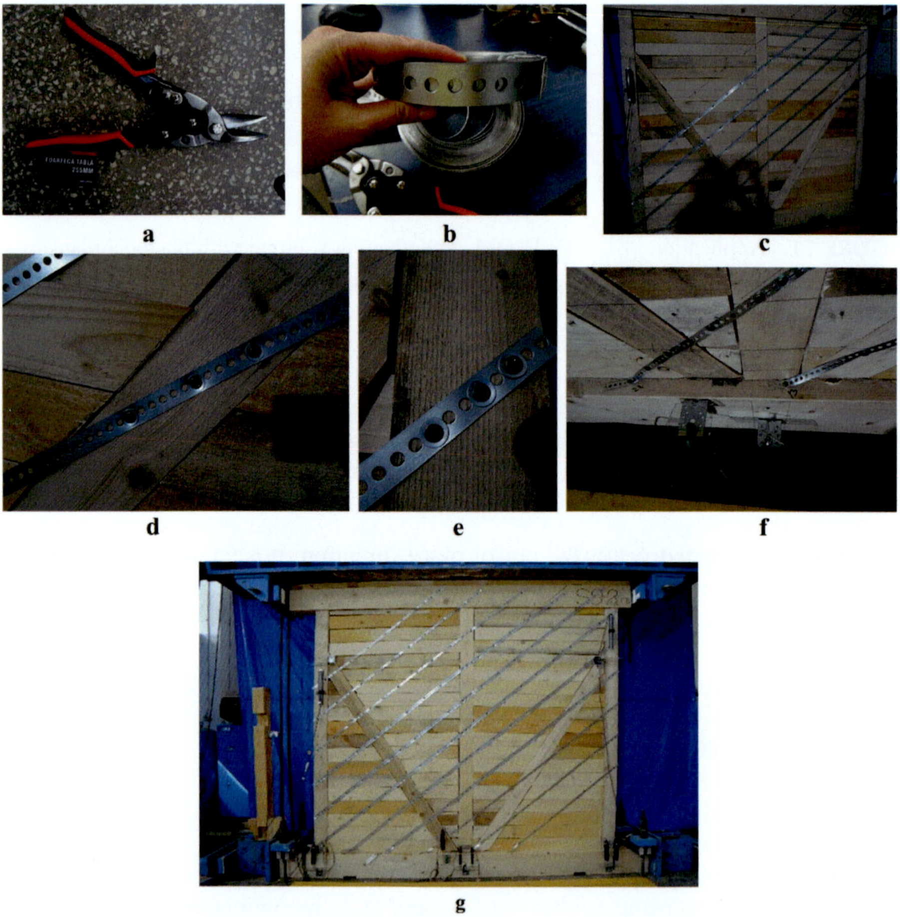

FIGURE 12 Application of the strengthening solution with fixing bands: (a) A sheet metal cutter is used to cut the band pieces to size; (b) the fixing band as bought from the shop; (c) the widest fixing band is applied starting from the main diagonal (the longest one) and continues on both sides of it, upward and downward; (d) on the main wooden frame, it is recommended to make at least two fastenings, preferably three; (e) washers are recommended for each screw; (f) in the sole plate area, the band is bent and secured by wrapping it around; (g) final layout of the specimen.

The solution is very clean and fast (if the screw drivers are working properly) and is especially suitable for the type 4 *paianta* (with a timber frame and horizontal timber logs infill) (Figure 4).

4.4.3 Dry Solution with Vertical Tie Rods

The solution involves the installation of pairs of vertical tie rods near the edges of the wall, positioned 30 cm inward from the wall's edges. A pair of tie rods consist of two upper tie rods, two lower tie rods, two central metal components that secure the tie rods (Figure 13a), and two turnbuckles (Figure 13b), which, by rotation, ensure the desired level of pretension. Besides these, at the bottom part and at the upper part of the wall steel jigs were designed and executed to act as load distributors to the wall. After connecting all the components, the tie rods are tensioned to a minimum of 2 kN each. At the upper part of the wall, between the frame and the masonry infills, due to the damages after the cyclic test on the non-strengthened wall (S4-a) [16], there was a gap that needed to be filled to complete the strengthening solution. Thus, timber planks were inserted to transfer the loads coming from the timber frame to the infills (Figure 13d).

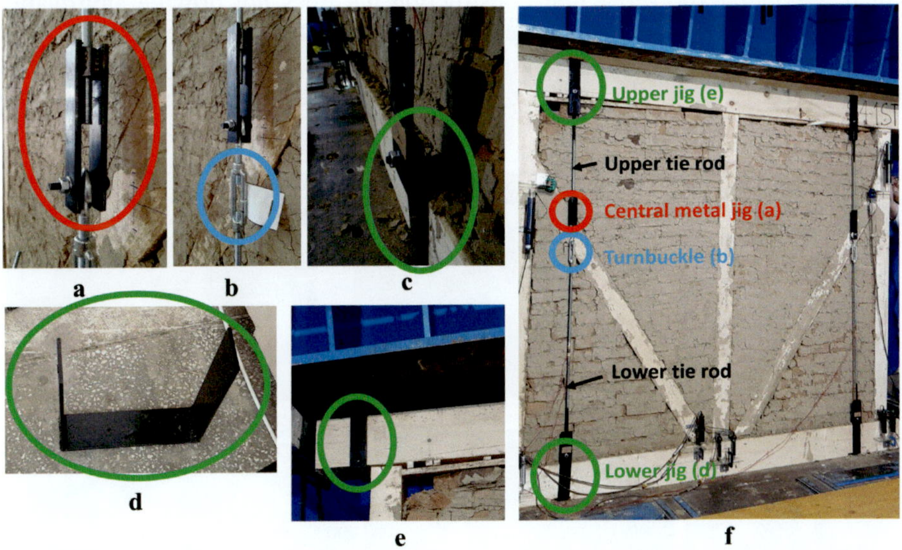

FIGURE 13 Application of the strengthening solution with tie rods: (a) The central metal component (jig) that connects the upper tie rod to the lower tie rod; (b) the turnbuckle, through which the pretensioning of the tie rods is achieved by rotating it, thereby tensioning the tie rods; (c) the tie rods are connected to jigs symmetrically in all four locations (2 on the front and 2 on the back); (d) the bottom "U-shaped" metal part (jig) which confine the wall in the vertical direction with the tie rods; (e) the upper "U-shaped" metal part (jig) which confines the wall in the vertical direction with the tie rods; (f) the final layout of the strengthening solution with tie rods.

4.4.4 Dry Solution with Metal Plates—S2-SMP (Wall without Infill) and S5-c-SMP (Wall with Infill)

This solution was tested on two walls, with and without infill, to observe the influence on the wall with/without infill. The used metal plates were L shaped for the bottom connections, and straight/plain plates for the connections between the braces and the posts at the upper part (Figure 14). The screws were 5x100 mm, the L shaped (corner) metal plates were 70x70x55x2 mm and the plain/flat metal plates were 55x135x2 mm. The envelope curve comparison of the specimens without infill strengthened vs. non-strengthened wall (S2-SMP vs.S2) is shown in Table 1.

The same metal plates application was done on a wall with timber frame and horizontal laths and mud and straw infill (type 3 *paianta*), and the results are shown also in Table 1, highlighting that the efficiency of this solution is greatly reduced for the infilled wall.

4.4.5 Dry Solution with Clamps—S3-SC6 and S3-SC8

A popular method to fasten connections in Romania in traditional carpentry, for both wall and roof framework is by using clamps, thus two types of clamps were applied on a wall without infill. The clamps were purchased from a local seller, which is popular for vernacular construction but does not provide any quality control, technical papers, etc. The clamps are basically reinforcement bars (usually used for concrete), smooth and ribbed, cut and bent to shape the clamp. Both bent parts were sharpened at their ends, so they can be inserted into the

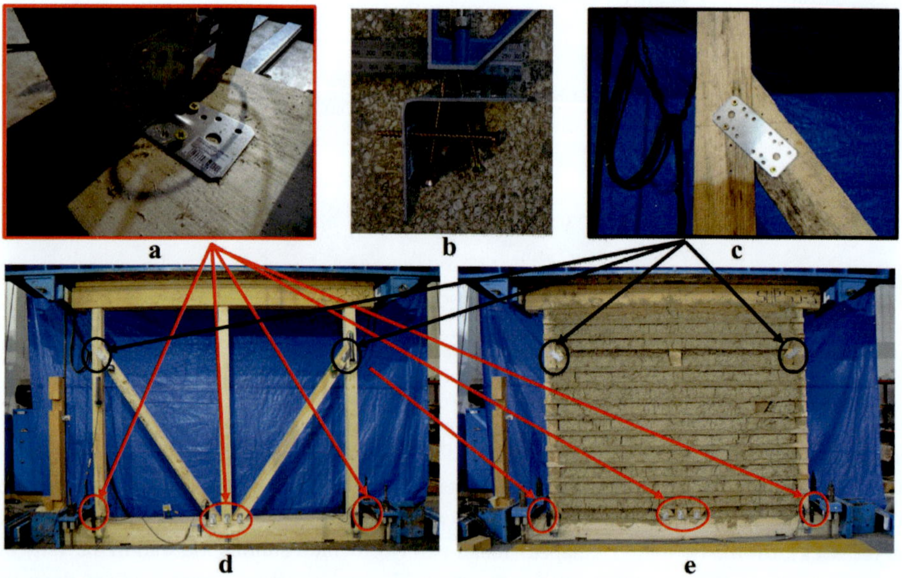

FIGURE 14 Application of metal plates L shaped (a, b), and plain (c) on a wall without infill, S2-SMP (d) and with infill (e) S5-c-SPM.

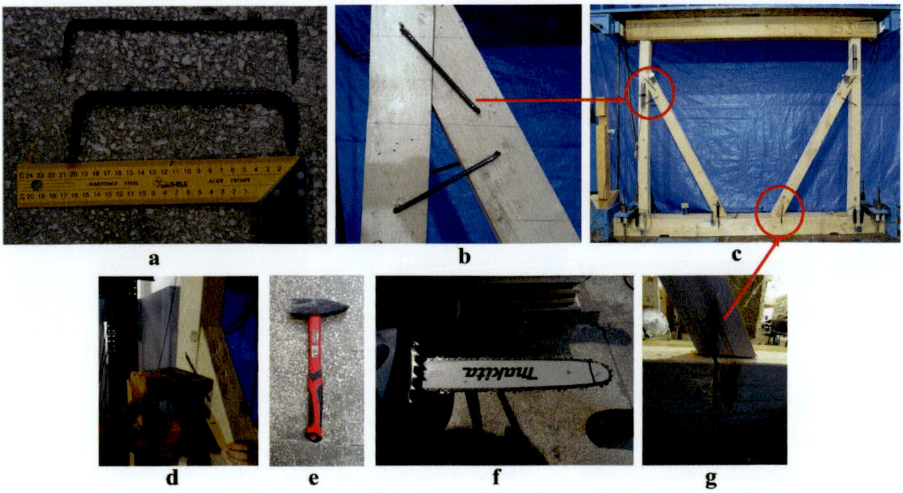

FIGURE 15 Clamp types (a); and their position on the specimen (b and c); a thick metal plate was used on the back side, when a clamp was inserted in the joint, to counteract the vibrations/movement of the wall when hammering the clamp into the position (d); the hammer used to insert the clamp (e); a small electric chainsaw (f) was used to make the pre-cuts at the bottom beam to be able to insert the clamp and accommodate the difference in cross-section between the bracing and the bottom beam simulating the mudsill (g).

timber easily using a hammer. Two diameters were used, for S3-SC6 smooth bar with diameter 6 mm, and for S3-SC8 ribbed bar with diameter 8 mm was used. The length of both types of clamps was 200 mm (Figure 15a). The clamps were hammered into position (Figure 15e), and on the other side, a counterweight was held by a second person, to overcome the balance and vibration induced in the wall when hammering the clamps into position. At the bottom part, due to the setup involving the bottom beam being thicker than the cross-section of the bracings, before hammering them, a slit was pre-executed into the bottom beam (Figure 15g) using a small electric chainsaw (Figure 15f).

The envelope curve obtained from the experiment, for the specimen with smooth clamps and ø6, and for specimen with ribbed clamps and ø8, respectively, versus the non-strengthened specimen is shown in Table 1.

4.4.6 Dry Solution with Punched Metal Plates—S5-c-SPMP

Punched metal plates are becoming popular for the roof framework nowadays in Romania, and they are available on the construction materials shops all over the country. Although the original punched metal plates need a special machine to insert them into the timber framework, it can also be done manually, with less quality, though. To understand the behavior of such plates, they were applied in the same locations as for the regular metal plates (Figure 16c): at the connection between the braces and the posts (Figure 16a), at the connection between the brace and the bottom

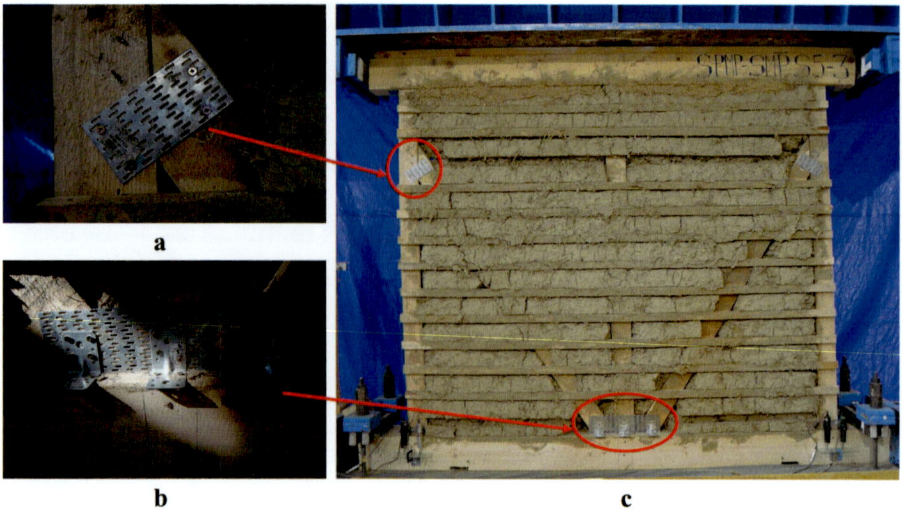

FIGURE 16 Application of the strengthening solution with punched metal plates (S5-c-SPMP): (a) At the connection between the brace and the post; (b) bottom connections between braces/intermediary post and bottom beam; (c) location of punched metal plates within the wall specimen.

beam (Figure 16b). For this latter part, due to the cross-section difference between the bottom beam and the brace, "L-shaped" metal plates were also used.

As can be seen from the envelopes' comparison to the non-strengthened wall (Table 1), the solution is ineffective. The reason is that after just a few cycles the punched metal plates fracture in shear and shred the timber elements they are attached to.

4.4.7 Dry Solution with Chains of Plastic Cable Ties—S5-a-SCT

Inspired by the PP-band investigated by Prof. Kimiko Meguro [20], plastic cable ties (Figure 17a) were connected as chains (Figure 17b) and the specimen was confined at different points (Figure 17e). The connection of the central vertical and diagonal cable tie chains was made at the bottom part, using ribbed clamps (ø8 diameter, 200 mm length) (Figure 17d) and for the horizontal chains, L-shaped metal plates were used (Figure 17c), attached to the timber frame with screw nails (ø6x80 mm). The envelope curve is compared to the one of the non-strengthened wall's in Table 1. The solution, although cheap and easy to apply, is very ineffective, due to the high flexibility of the chains.

4.4.8 Dry Solution with Used Tires Bands—S6-a-STY

This solution was inspired from the fact that every car owner in Romania, at some point must change the car tires, and the old ones are just stored somewhere and occupy the space for nothing or thrown away as garbage. One way to recycle, would

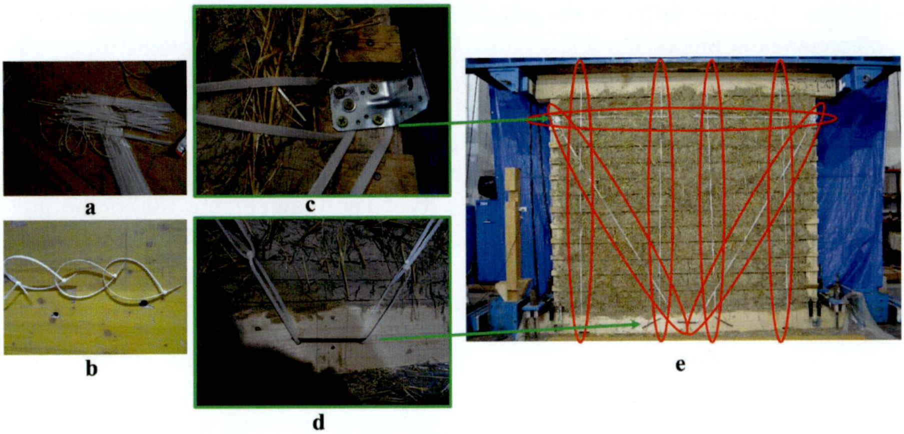

FIGURE 17 Application of the strengthening solution with plastic cable ties: (a) How the cable ties come in the package; (b) creating chains of cable ties; (c) connecting the horizontal and diagonal chains to the timber frame, by using corner metal plates and screw nails with washers; (d) at the bottom part, the clamps used to connect the diagonal cable tie chains; (e) the locations of the chains on the wall specimen.

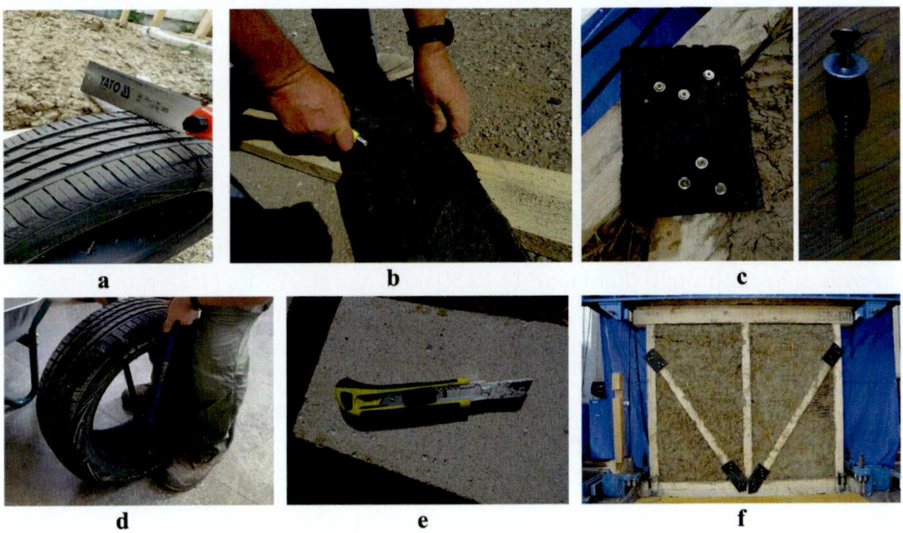

FIGURE 18 Application of the strengthening solution with used car tire bands: (a) trial to cut the tire with the Ryoba pull saw; (b) the winner solution was to use a simple cutter to create the bands; (c) washers and screw nails used to connect the bands to the timber frame; (d) using a bolt cutter was unsuccessful; (e) the only way to cut a tire is with the classic and simple cutter; (f) the locations of the car tire bands placed on the strengthened wall.

be to reuse them for different things, and for this purpose, old car tires were cut to small rectangular pieces and screw nailed (same ones used before, ø6x80 mm) to the parts of walls that showed the most damage/deformation, as an attempt to reinforce the connections between the bracings and the timber frame, and provide a material that can elongate without fracture, to dissipate the seismic energy.

One challenge was to cut the bands to the required dimensions. Even if we tried to cut the tire with a bolt cutter (Figure 18d), a Ryoba pull saw (Figure 18a) or the angle (disc) grinder, none of these solutions worked, except the classic and simple cutter (Figure 18b, e).

Washers were used for all the screw nails to prevent the fracture punching of the car tire bands in the execution phase (Figure 18c). The locations of the car tire pieces are shown in Figure 18f.

The envelope comparison with the non-strengthened specimen is shown in Table 1. This result shows that the solution is not efficient at all, as there is no increase in stiffness, nor in strength. This may be due to the flexibility of the tire bands, which proved that it is not helping the timber frame.

4.4.9 Wet Solution with Welded Mesh and Cement Mortar

This solution is very popular among the non-engineering procedures for traditional houses nowadays. People in the rural area often do not ask for engineering advice when they want to retrofit their traditional house, and this is one easy solution, for which the materials are widely available on the market. Thus, to reproduce it in the lab, the welded wire mesh was 4 mm thick, with meshes of 100x100 mm. The mortar's recipe was 2:6 (cement: sand) and it was executed as follows: one 100 l wheelbarrow of sand, half of a cement bag (type Carpatcement CEM II/A 42.5R, 40 kg), half a bucket of water (the bucket's capacity was 12 l). Compression tests were done on mortar cylinders, and the average compression stress obtained was 21.6 MPa. The mortar layer was about 20 mm thick. The connection between the wire mesh and the wall was done with small plain nails (1.8x50 mm) and steel wire (1.18 mm diameter), the quantity was about 6 pieces at every m^2. The thickness of the mortar was 2 cm on each side. It took three days for two people to execute; it finished on October 13th and was tested on November 14th.

The wall that was strengthened with this solution was a timber frame infilled with autoclaved aerated concrete (AAC) brick masonry. The steps to apply the strengthening solution are shown and described in Figure 19a to Figure 19g.

4.4.10 Wet Solution with Rabbit Mesh and Cement Mortar

The wall that was retrofitted with this solution was a timber frame infilled with burned clay brick masonry with earth mortar, which was not tested before the application of the strengthening solution. This is another popular alternative to the steel welded wire mesh solution presented previously, where the rabbit wire mesh for plaster is used to reinforce the cement mortar applied on the wall. The rabbit wire mesh is made of zinc coated steel wire, braided to form hexagon meshes, and in our case had a diameter of 0.8 mm. The mortar recipe was the same as for the previous solution.

FIGURE 19 Application of the strengthening solution with steel welded wire mesh and cement mortar (S8-PS): (a) applying the welded wire mesh in this case it was 4x100x100 mm, and profiled (not smooth); (b)connecting the wire mesh with nails inserted in the timber frame of the wall by means of a thin and smooth steel wire (ø1.18 diameter); (c) applying a first mortar layer carefully between the wire mesh and the existing wall (less than 1 cm distance between them); (d) adding the mortar with the target thickness on top of the timber posts, to serve as a guide for the rest of the mortar layer; (e) the rope attached with nails on an additional wooden plank attached to the marginal posts, to serve as a guide for the final thickness of the mortar layer; (f) the mortar was applied in lumps, to fill in all the gaps, with a special attention to the areas between the timber frame and the infills; (g) the final layout of the strengthened wall.

The thickness of the mortar layer was also about 20 mm, with an approximate distance of up to 1 cm between the wire mesh and the existing wall. Compression tests were done on mortar cylinders, and the average compression stress obtained was 31.5 MPa. The connection between the wire mesh and the wall was done with screw nails (1.8x50 mm) and steel wire (1.18 mm diameter), the quantity was about six pieces at every m^2. Two people finished it in two days (eight hours working day). The steps to apply the strengthening solution are shown and described in Figure 20a to Figure 20h.

4.4.11 Wet Solution with Plastic Mesh and Earth Mortar

The wall that was strengthened with plastic mesh and earth/clay mortar was made of a timber frame with wattle and daub infill. The plastic mesh was chosen as an alternative to the classic steel/rabbit wire meshes, considering that plastic does not corrode in contact with the earth/clay mortar as these classic options do. The plastic mesh was selected randomly from the local market, and the clay mortar was made of earth and sand (4:1), and mixed with straws to prevent cracking during the drying phase (Figure 21d). Professional construction-grade plastic meshes were intentionally avoided due to their high cost. The connection of the plastic mesh to the wall, on the infill area, was done with steel wire (Figure 21a), passing through the depth

FIGURE 20 Application of the strengthening solution with rabbit wire mesh and cement mortar (S10): (a) Applying the welded wire mesh in this case it was 0.8 mm diameter; (b) connecting the wire mesh with screw nails inserted in the timber frame of the wall by means of a thin and smooth steel wire (ø1.18 diameter); (c) using a drilling machine with a long drill to make holes from one side to the other in the infills; (d) in the drilled holes, steel wire was inserted and knotted on one side, to create a connection between the meshes on both sides of the wall and the infill; (e) mixing the mortar directly in the wheelbarrow with an electrical hand concrete mixer; (f) the mortar was applied with trowels; (g) applying a first mortar layer carefully between the wire mesh and the existing wall (less than 1 cm distance between them); (h) the final layout of the strengthened wall.

of the wall (through pre-drilled holes), so the wire can connect the plastic mesh on both sides of the wall. On the timber frame area, the mesh was connected with nails (Figure 21b). Before the mortar application, the existing surface was sprayed with water (Figure 21c). The mesh sheets, which had a 1 m width, were overlapped on 30 cm (Figure 21e). The mortar is applied in chunks on the wall (Figure 21f), and leveled with trowels (Figure 21g). The wall was kept in the lab to dry for at least two months and final layout is shown in (Figure 21h).

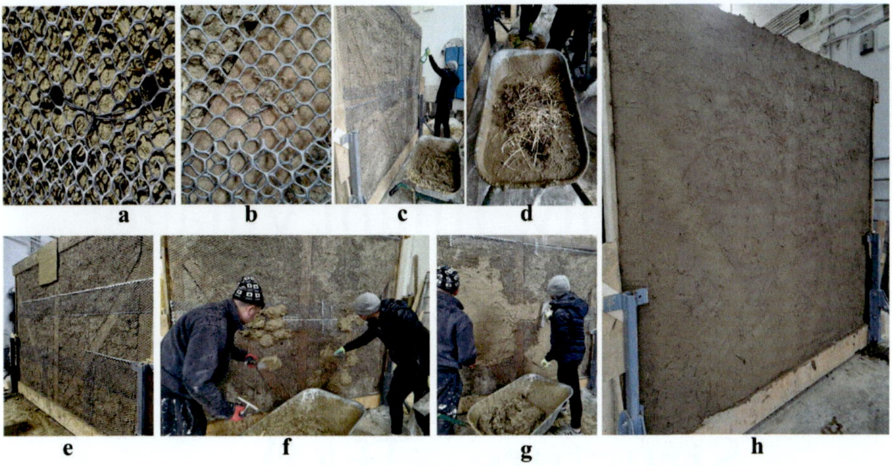

FIGURE 21 Application of the strengthening solution with plastic mesh and clay/earth mortar (S6-b-PMClay): (a) Applying the plastic mesh connecting it to the existing wall by means of a thin and smooth steel wire (ø1.18 diameter); (b) connecting the plastic mesh to the timber frame plain nails inserted in the timber frame of the wall; (c) using a drilling machine with a long drill to make holes from one side to the other in the infills; (d) in the drilled holes, steel wire was inserted and knotted on one side, to create a connection between the meshes on both sides of the wall and the infill; (e) mixing the mortar directly in the wheelbarrow with an electrical hand concrete mixer; (f) the mortar was applied with trowels; (g) applying a first mortar layer carefully between the wire mesh and the existing wall (less than 1 cm distance between them); (h) the final layout of the strengthened wall.

5 Case Study (Single-Family House in Prahova County, Tufani Village: Timber Frame with Brick Masonry Infill Structure—TFM)

In this chapter, a vernacular house (Figure 22) was investigated from the structural engineer's point of view, and all the details are further presented. The house represents a classic *paianta* house, with timber framed walls and clay brick masonry infills. Usually, such houses have only ground floors. The ground floor plan is shown in Figure 23.

5.1 SITE CHARACTERISTICS

The seismicity of the area, in accordance with P100-1/2013, which is the National Design standard in Romania, is characterized by the peak value of the design ground acceleration ag = 0.35g (with IMR = 225 years and 20% probability of exceedance in 50 years) and the corner period Tc = 1.6 s.

5.2 DESCRIPTION OF THE CONSTRUCTION

5.2.1 Geometric Characteristics

The building, with the function of a single-family dwelling, is developed within a single level (P—ground level) and has the geometric characteristics that are shown in Figure 24, Figure 25 and Figure 26, and summarized as follows:
In the horizontal plane:
Dimensions of the footprint on the ground:

- Maximum width at the footprint on the ground: 4.965 m.
- Maximum length at the footprint on the ground: 12.38 m.

The distances between the axes of the construction in the longitudinal direction are:

DOI: 10.1201/9781003647249-5

FIGURE 22 Single-family home with a ground floor height regime, with the wall structure made up partly of braced wooden frames with masonry infill, partly of simple masonry.

- 1.275 m between axes A and B.
- 1.77 m between axes B and C.
- 1.595 m between C and D.

The distances between the axes of the construction in the transverse direction are:

- 3.095 m between axes 1 and 2.
- 1.00 m between axes 2 and 3.
- 2.515 m between 3 and 4.
- 5.445 m between 4 and 5.

Vertically:

- The height from the elevation of the developed land to the upper part of the floor at elevation ± 0.00 is 0.565 m.
- The height from elevation ± 0.00 to the eaves is +2.50 m.

5.2.2 Description of the Structural System

The **infrastructure** of the building consists only of continuous plinths made of unreinforced concrete (lower class) arranged under the walls of the building, having a height of about 50 cm, and the width is about 30 ÷ 35 cm for walls with a thickness of half a brick and about 40 cm for walls with a thickness of one brick. The **plinth**

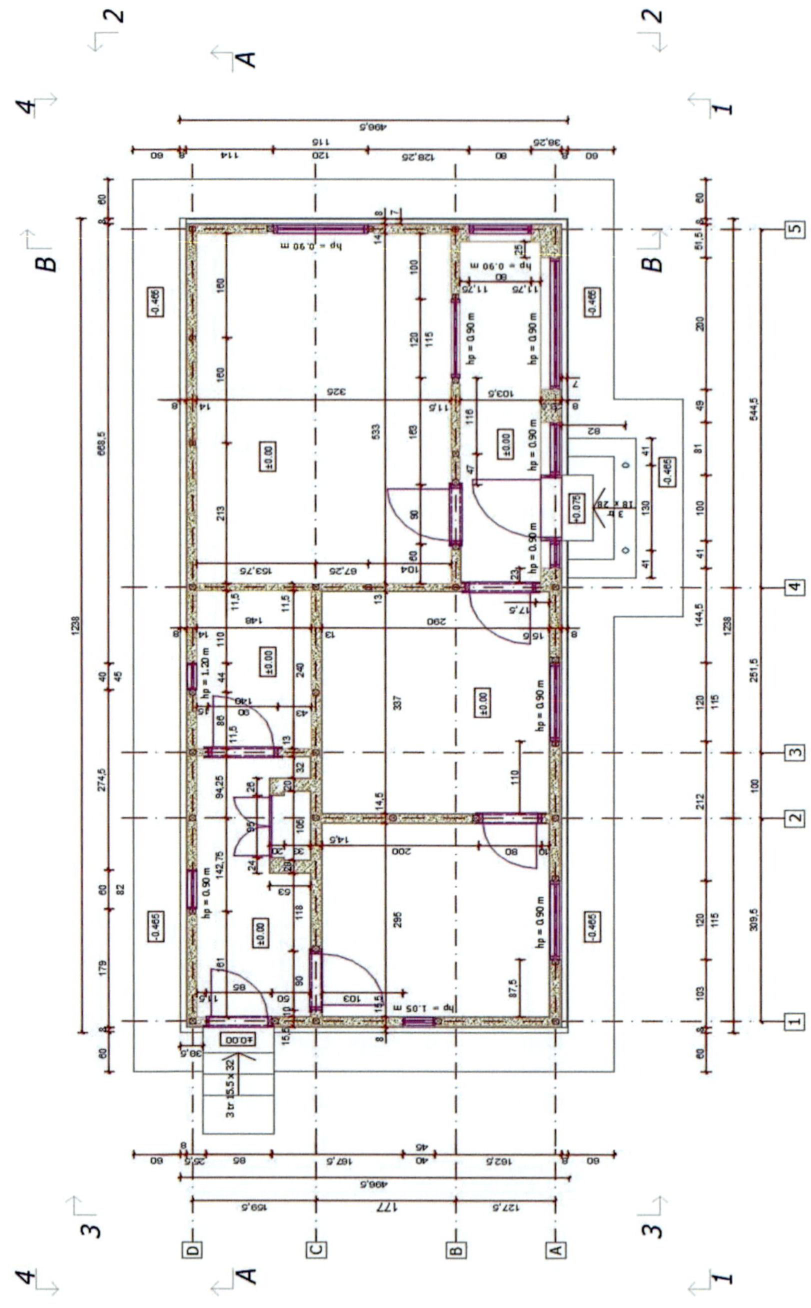

FIGURE 23 The floor plan.

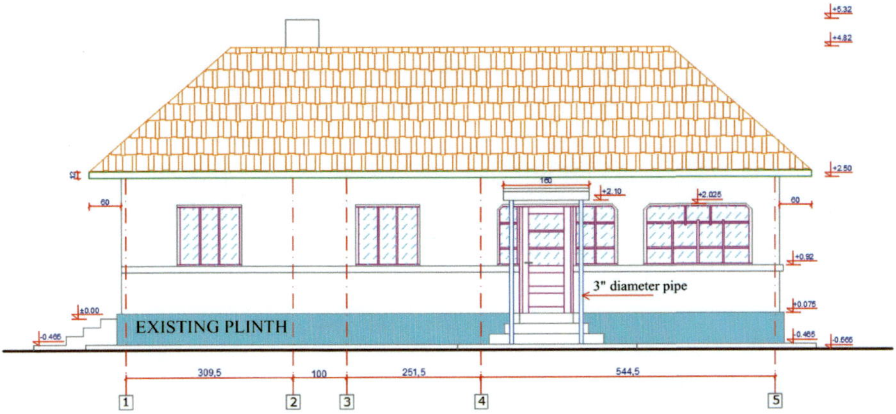

FIGURE 24 Frontal view of the main entrance (showing the plinth).

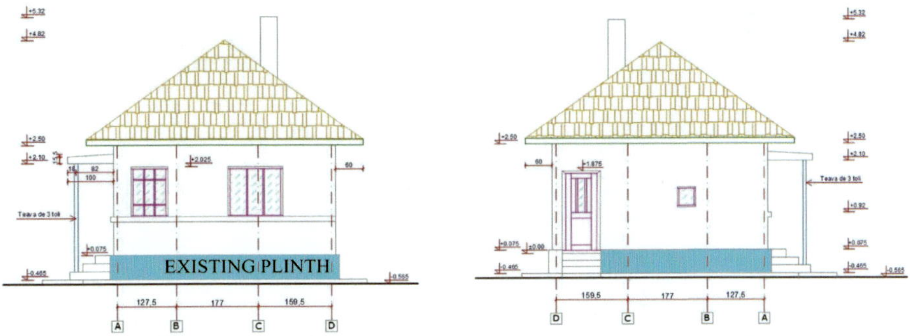

FIGURE 25 Lateral view (right side of the main entrance) (left) and Lateral view (left side of the main entrance) (right).

in a building is the part of the structure located **between the ground level and the finished floor level (FFL) of the ground floor**. This house is lacking a foundation.

The wooden posts that make up the walls' framework, especially at the corners of the house, pass through the plinth, being driven into the ground about 20 cm. In the space bordered by the plinths, a filling of compacted earth is made in a thickness of about 35 cm, over which there is a layer of gravel, against capillary action of water, in a thickness of about 16 cm (Figure 27).

The **stairs** that provide access to the building (between the courtyard level and the level ± 0.00) are made of plain (unreinforced) concrete. The **perimeter sidewalk** is approximately 60 cm wide, being also made of plain (unreinforced) concrete.

In the entire superstructure of the building, the **load-bearing walls** are arranged both in the transverse and longitudinal directions, and they are made in two construction variants. The envelope walls located in axis A and between axes 4 and 5, respectively in axis 5 and between axes A and B, are made of unreinforced clay

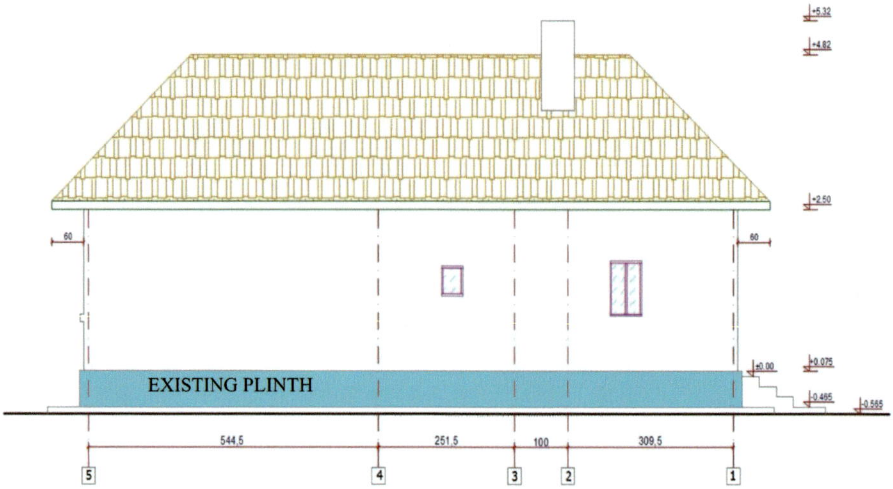

FIGURE 26 Back side view.

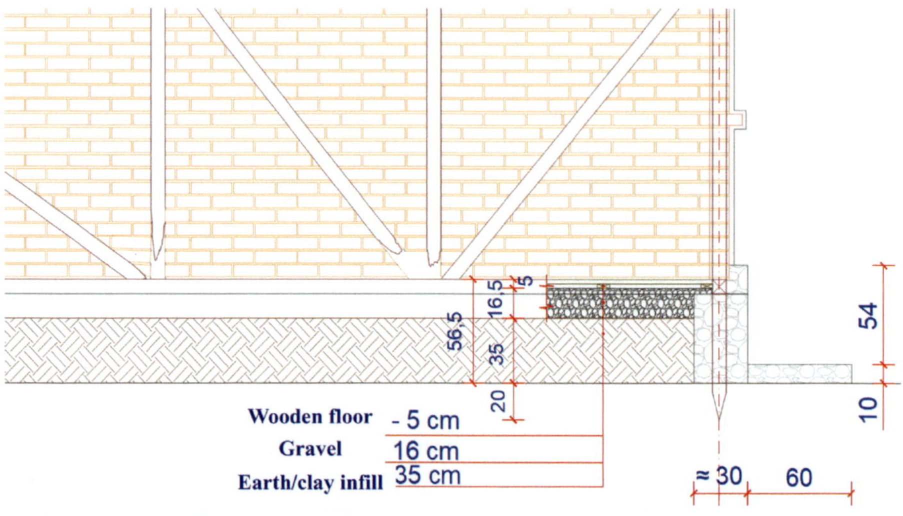

FIGURE 27 Detail regarding the wooden posts passing through the plinth (at the intersection of the walls) and the layers executed in the space bounded by the plinths [cm].

brick masonry (240 x 115 x 63 mm), having a thickness of one brick (Figure 28a). Otherwise, the walls of the building (located on both sides) have a mixed structure, composed of braced timber frames (made of squared wooden pieces with a section of approximately 8 x 8 cm) and solid clay brick masonry infills (240 x 115 x 63 mm) with a thickness of half a brick (Figure 28a). The layout of the axis D wall is shown in Figure 29.

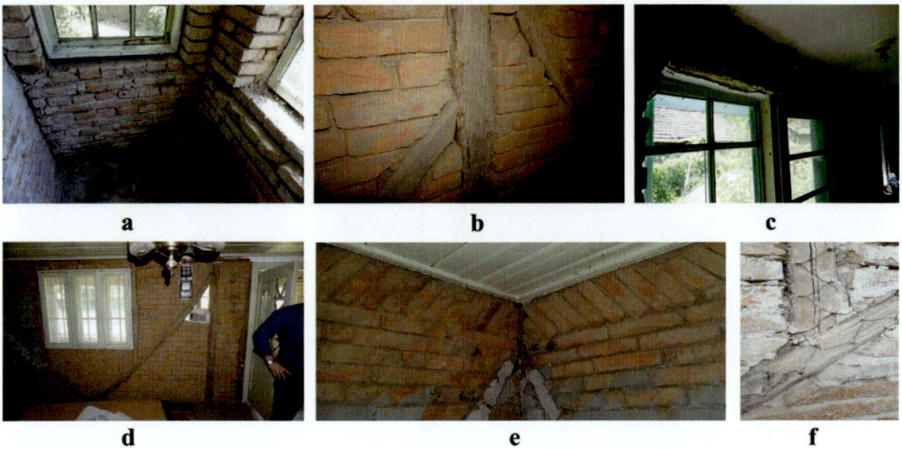

FIGURE 28 Structural observations of the house: (a) Single-brick masonry envelope walls (which border the entrance hall from the outside); (b) Composite walls, with a frame structure made of squared timber pieces, braced and filled with clay brick masonry; (c) Wooden lintels located above the window openings, over which several masonry piers were partially built, partially a concrete portion; (d) Lack of lintels above window openings in mixed-structure walls; (e) The upper portion of the masonry panels (non-woven elements) in the composition of the walls with mixed structure (wooden frames with masonry infill) (exposed area); (f) wire connected to nails inserted in the timber frame, to ensure the support of the plaster on the wall.

The timber frames in the structure of the mentioned walls are made up of posts arranged at the ends and in some intermediate locations, top plates that rest on the upper part of the posts (but also on the filling masonry), mudsills arranged between the marginal posts (resting on the plinth) and inclined/diagonal bracing elements that connect the mudsills with some posts. The height of the wall system, including the thickness of the upper floor, is approximately 2.40 m.

Above the window openings, located in the one-brick thick walls (which border the entrance hall from the outside), wooden lintels (which show deterioration) were observed, over which three courses of masonry are laid. Above the access door to the hall, over the wooden lintel (Figure 10), a reinforced concrete (RC) beam (24 x 22 cm) was executed, over a length of 1.60 m, in which a RC slab with an opening of 1.00 m and variable thickness (h_{slab} = 15 ÷ 20 cm) is embedded (Figure 28c), supported on the opposite edge of the wall by two Ø80 mm steel pipe posts that flank the access staircase. Above the rest of the window openings (made in the walls with mixed structure) no lintels were observed, the masonry discharging directly onto the window frames (Figure 28d). In the case of TFM (timber framed masonry) walls, the infill masonry is woven to the height of the panels, except for the last portion under the upper sole, where the ceramic elements are arranged in an inclined position, not being woven (Figure 28e). The walls of the building are plastered with mortar, with a thickness of about 1.5 cm on the inside and 2.5 cm on the outside. To ensure adhesion between the mortar and the wooden parts of the wall structure, nails were hammered into them, and wire connections were made (Figure 28f).

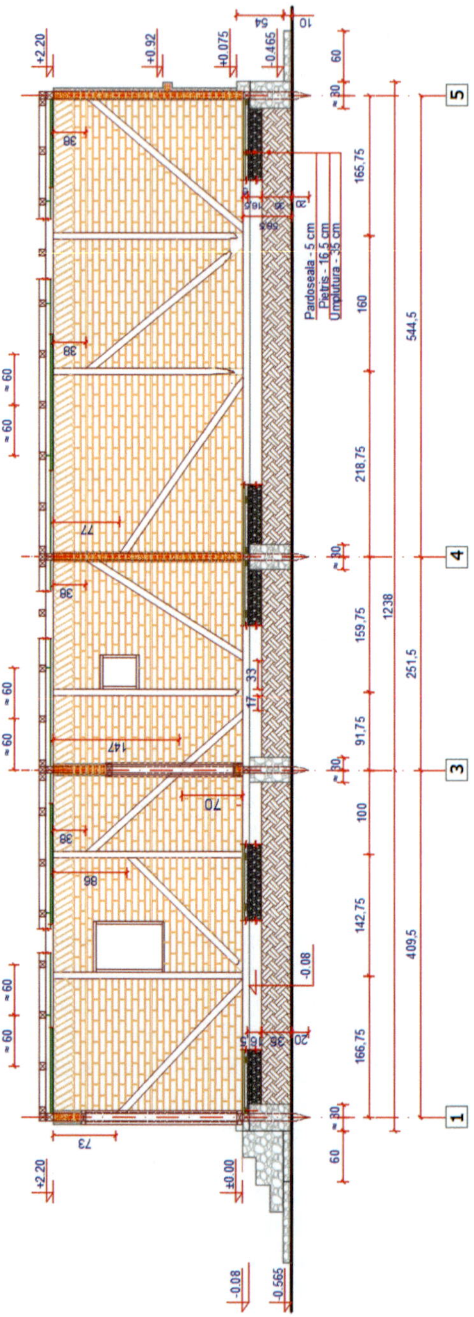

FIGURE 29 Axis D wall (without finishing).

At the top, the structural wall system is "connected" by a slab made of wooden beams oriented in the short direction of the building (spaced about 60 cm) and a timber plank floor. Immediately below the slab beams, perpendicular to their direction, a network of tangent elements is fixed, made of raw processed wood pieces (with a section of ≈ 7 x 5 cm) (Figure 30a), which constitutes the ceiling structure, respectively which supports the thermal insulation layer (inserted between the slab beams) made of a rubble infill (mixed with clay/earth).

On the soffit of the slab above the entrance hall (the portion between axes 4–5/A – B in Figures 23 and 31), the ceiling is made of plaster reinforced with rabbit mesh, fixed to the rough-processed timber pieces, located immediately below the slab beams (Figure 30b). The rest of the ceiling is made of wooden paneling (about 20 cm wide) fixed with nails to the network of rough-processed tangent elements (Figure 16).

The **roof framework** is a four-angled structure, made of squared and round timber elements, joined with the help of staples and nails. The rafters of the framework are made of round wood with $\phi \approx 12$ cm, spaced about 90 ÷100 cm apart. Slats with a section of ≈ 3.8 x 5.8 cm, spaced about 25 ÷ 30 cm apart, rest on the rafters, which support the roof tile covering. The roof tiles are ceramic, measuring about 40 x 22 cm (Cărpiniş Tile Factory) (Figure 30d).

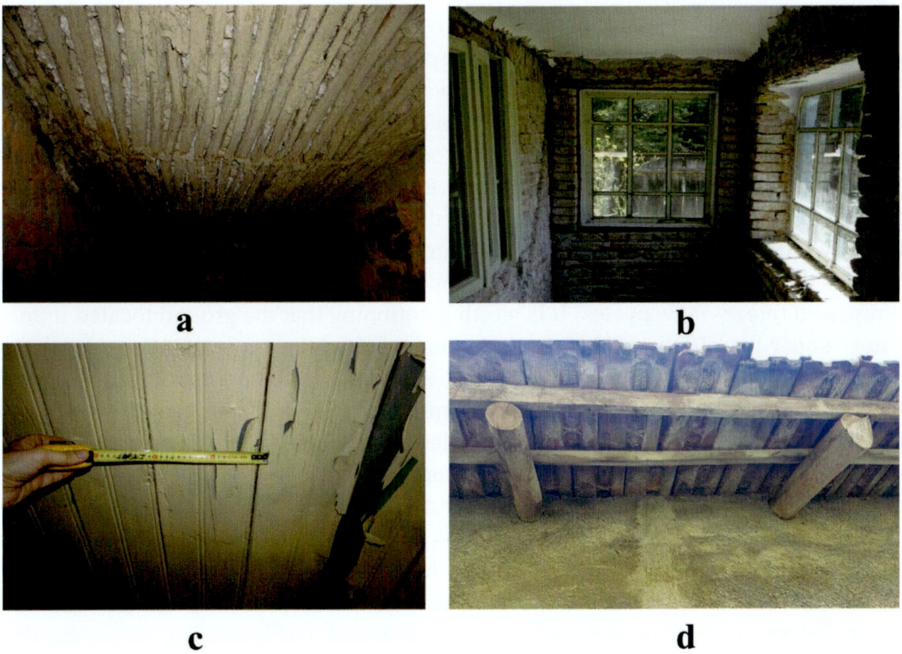

FIGURE 30 (a) The soffit of the slab above the ground floor. The network of processed raw wood pieces fixed to the soffit of the beams can be distinguished; (b) The ceiling located on the underside/soffit of the slab above the entrance hall; (c) (Timber planks fixed to the network of roughly processed tangent elements, forming the ceiling; (d) The roof cover, resting on the network of battens that unload on the rafters.

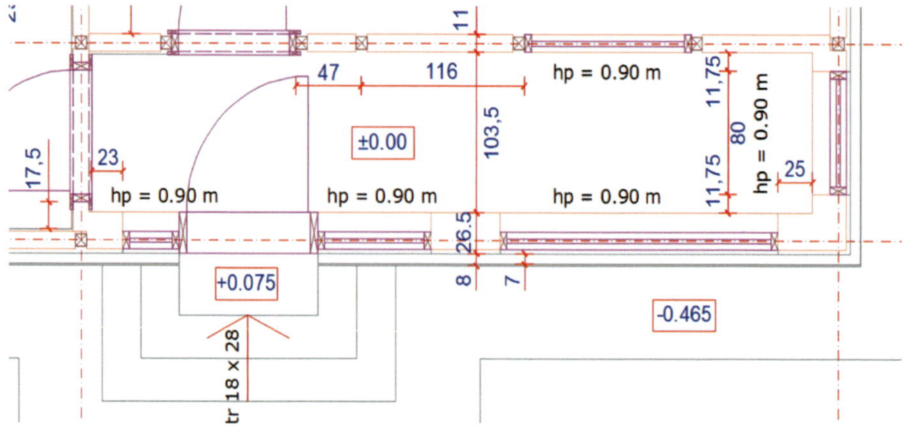

FIGURE 31 Plan view of the entrance hall.

5.3 OBSERVED DEGRADATIONS IN THE CONSTRUCTION

5.3.1 Damages in the Superstructure

Damaged, displaced, or missing tiles were observed on the **roof cover**. At the same time, the roof does not benefit from a system of gutters and downpipes (Figure 32) that would allow the evacuation of rainwater away from the base of the walls, an aspect that allows both infiltration and deformation of the foundation soil under the base.

Cracks with a slightly inclined orientation to the vertical, and openings of about 1.5–2 mm, were observed in the envelope walls on the main entrance façade (axis A; Figures 23 and 32b) and the street-facing side (axis 5; Figure 32c). These cracks initiate at window corners and propagate toward the sidewalk level.The cause of these cracks is the subsidence produced in the ground under the plinth of the walls, because of repeated freeze–thaw cycles. It is worth mentioning that the ground located immediately under the plinths of the envelope walls, being close to the external environment, is strongly exposed to the action of atmospheric temperature, which during winter registers negative values. At the same time, the production of the cracks in question was also facilitated by the cancellation of the tension effect of the mudsill on which the masonry was executed, a situation caused by the local rotting of these timber elements, exposed to moisture migrated from the ground (the waterproofing system between the mudsill and the plinth is lacking).

The cracks (2 ÷ 3 mm), recorded at the level of the envelope wall in axis A (the wall containing the main access door) can also be observed from inside the building and in the vicinity of the intersection of axes A and 1 (Figure 33), an aspect that indicates a penetration of the wall (Figure 32d). This fact clearly highlights the state of degradation of the base of the wall.

Cracks were also observed at the level of the envelope wall in axis 1, extending along the entire height of the mentioned wall, intersecting the window opening (Figure 34a). The cracking of the walls, as a result of the settlements recorded under

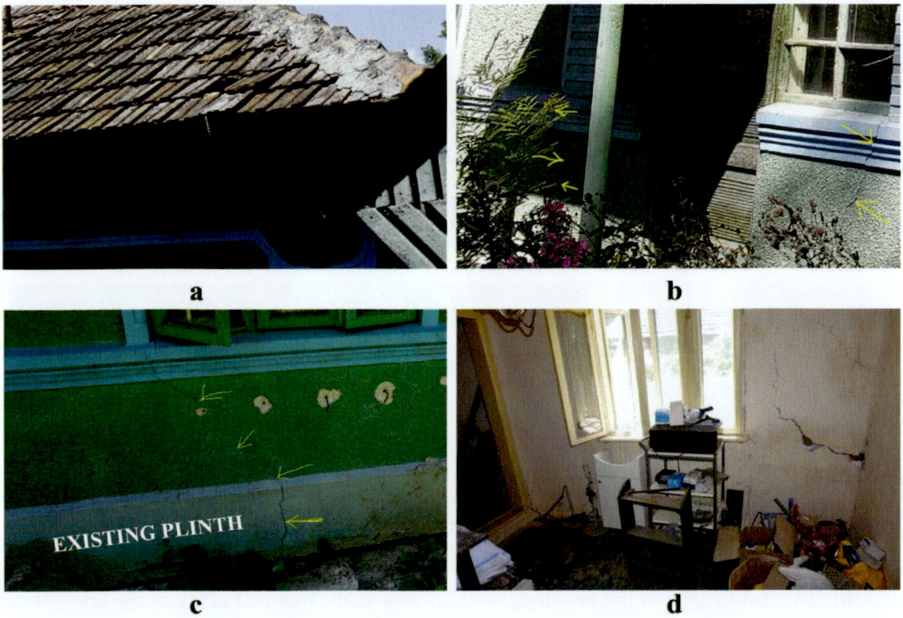

FIGURE 32 (a) Degraded roof tiles and lack of gutter system; (b) cracks in the tire wall located in the A-axis, in the immediate vicinity of the access door; (c) crack in the envelope wall located in axis 5 (from the street). The crack starts at the corner of the window and goes towards the sidewalk; (d) crack in the wall of axis A, immediately near its intersection with the wall of the envelope of axis 1.

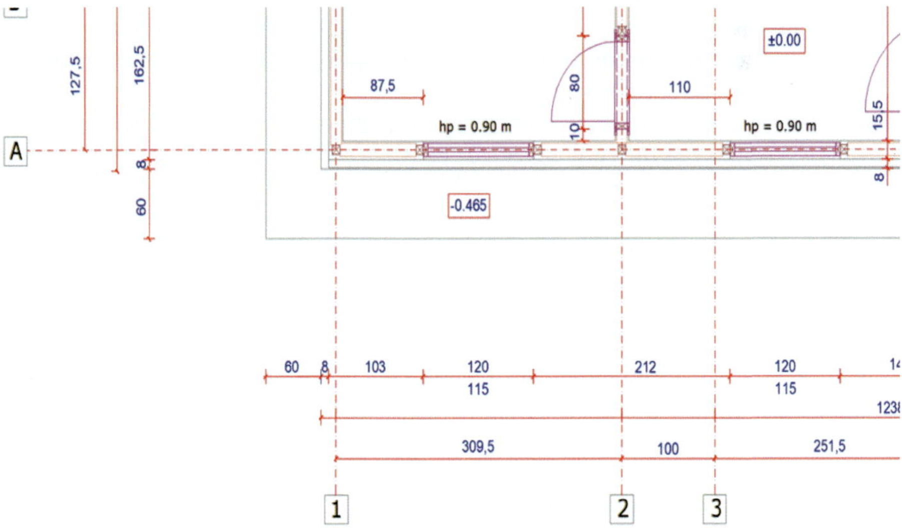

FIGURE 33 In plain view of the cracked area at the level of the envelope wall in axis A, in the vicinity of the corner located at the intersection of axes 1 and A.

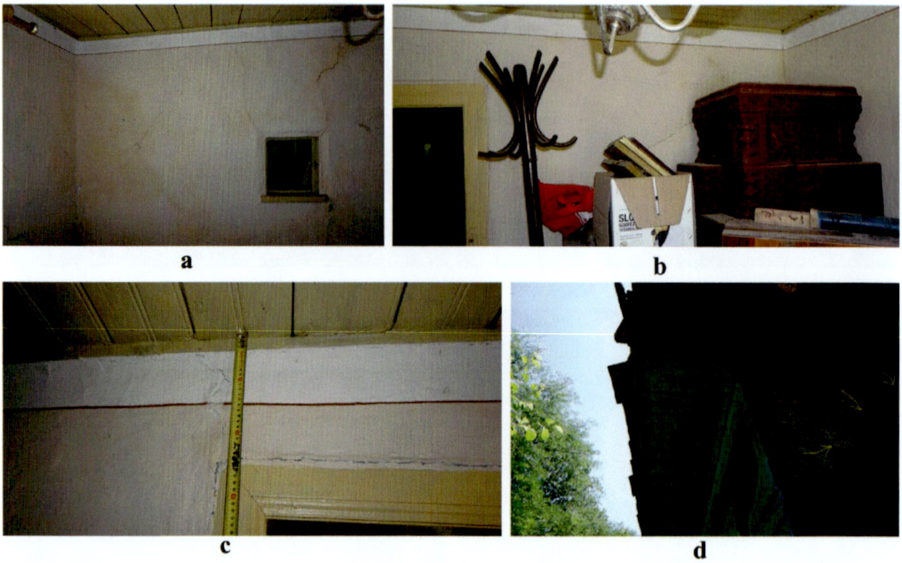

FIGURE 34 (a) Cracks in the envelope wall on axis 1; (b) Inclined crack at the level of the inner wall in the C axis; (c) Vertical crack in the interior wall of axis 2, immediately above the passage door; (d) Crack at the top of the envelope wall located in axis D (behind the house).

the plinths, can also be observed inside the building, more precisely at the level of the wall in axis C (to the right of the passage door) (Figures 24, 34b and 35), as well as at the level of the wall in axis 2, above the door frame (Figure 34c). Also, cracks with an opening between 0.5 ÷1.0 mm were observed, with an inclined path, at the upper part of the envelope wall located in axis D (the wall behind the house), near the corner located at the intersection of axes 5 and D. It is possible that the crack is located only at the plaster level (thus superficial). The cause of these cracks is the combined effect of possible water infiltrations into the structure of the mentioned wall (because of local degradation of the roof cover), or possible settlements at the foundations or thrusts coming from the roof framework (Figure 34d).

In some locations at the bottom part of the envelope wall in axis D (backside of the house), portions were observed where the plaster is fallen off and the concrete of the plinth is heavily degraded (cracks, significant exfoliation, the concrete tends to crumble), location where the timber elements of the wall structure (posts and mudsill) affected by rotting are observed, their sections being interrupted. Also in these areas, the wall's masonry is locally loosened, the bricks being displaced (Figure 36a, b). The damage observed on the outside of the envelope wall in axis D is also visible from the inside of the building, being developed throughout the entire thickness of the wall (Figure 36c).

Note. The lack of a foundation system under the walls of the building, combined with the lack of waterproofing measures for the bottom of the walls or the ineffective measures adopted in this regard during the construction of the building, led over time to deformations in the wall structure, therefore to additional stress states, as well as

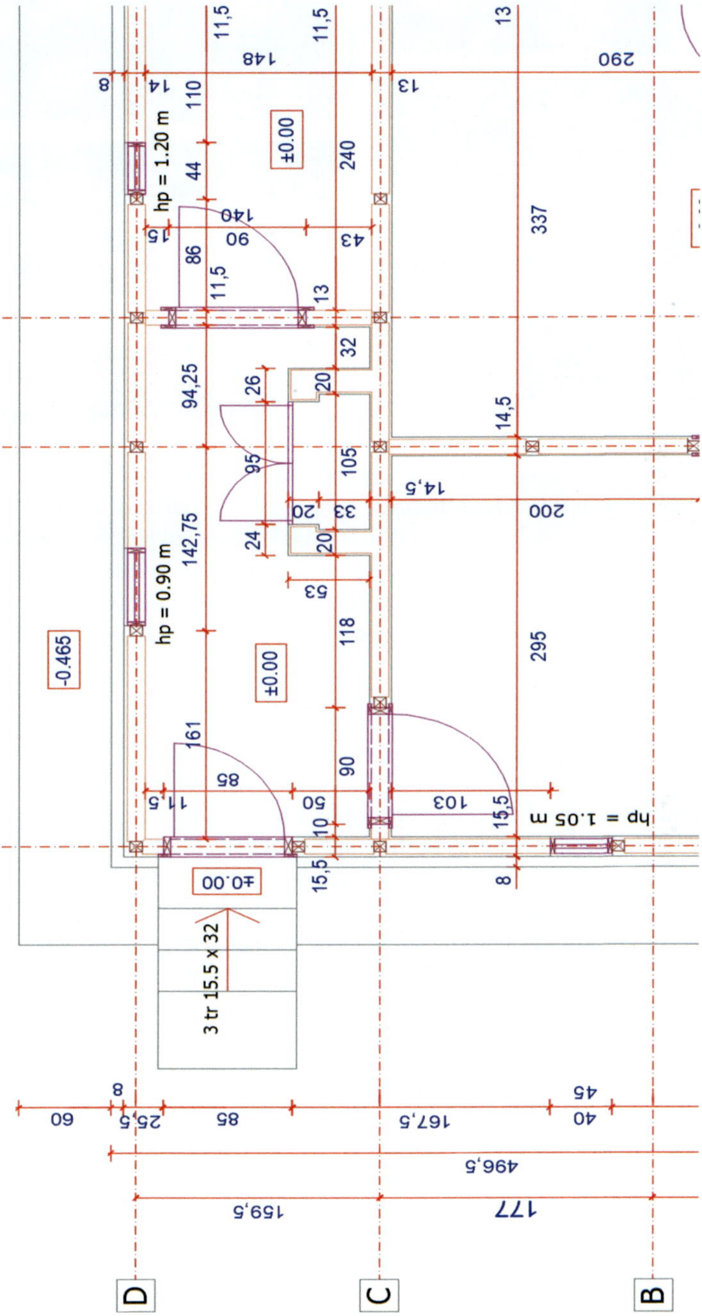

FIGURE 35 Wall in axis C.

FIGURE 36 (a) Masonry failure in areas where sections of timber elements embedded in the envelope wall's structure in axis D have disappeared because of the effect of rotting; (b) Rotting damage at the base of the envelope wall from the D-axis. The affected plaster and masonry can be observed; (c) Damage due to rotting of the wooden elements embedded in the structure of the envelope wall at the back of the house (axis D). The affected masonry is visible; (d) The outer surface of the plinth shows portions of exfoliated concrete; (e) the concrete in the steps of the access staircase is also damaged.

to infiltrations that spread over time to larger areas, their combined effect causing the aforementioned degradations.

The structural concept of the envelope walls that border the entrance hall is not redundant, due to the large openings of the functional gaps (2.00 m and 2.22 m) made in the wall. These (with a thickness of one brick) being in the alignment of axis A and between axes 4 and 5 delimit the access hall – over which timber lintels are arranged (not anchored to the masonry and which have damaged portions). Moreover, the masonry parts located between the openings and the corner area, have short lengths (23, 49, and 51 cm) and lead to H/L ratios (between their height and length). All these expose their unconfined sections to combined stresses (τ and σ) with potential harmful effects under the action of lateral seismic forces, the composition of the envelope walls that border the entrance hall is not redundant.

5.3.2 Degradations in the Infrastructure

At the level of the plain concrete plinths located under the envelope walls of the building, where the plaster has fallen, significant degradation (ruptures, areas with crumbly concrete, and exfoliated portions) was observed, as a result of repeated exposure to freeze–thaw cycles in a wet state, respectively of the movements produced in the ground (Figure 36d).

Given the fact that the porosity of the concrete in the plinth is high (as a result of deficiencies in preparation and compaction during execution), as well as the lack of waterproofing at its level, the probability that the aforementioned state of degradation will be found in other areas, including under the interior walls, is high.

Exfoliation of the concrete was observed also at the level of the access staircase in the building, the facade in axis A, as a result of repeated freeze–thaw cycles (in a wet state) (Figure 36e).

5.4 ASPECTS REGARDING THE BUILDING'S STRENGTHENING

5.4.1 General Considerations

Interventions for the strengthening of any existing house should be carried out based on a strengthening project, prepared by a structural engineer, which in turn is drawn up based on the conclusions provided by the previously prepared technical expertise report, which in Romania is a technical evaluation of the structural system done by a national certified expert. The Ministry of Construction provides an updated list with all the certified experts in the country from which the owners can choose and can contact them. The experts are structural engineers who passed a national examination, and thus are certified.

When developing the project, possibilities on how to improve the structural system (introduction of new elements in conjunction with the original ones, new supports and connections, etc.) will also be analyzed, an aspect that can simplify the intervention and also improve the costs of the work.

The execution of the works can only begin after the construction has been unloaded (of everything possible) and the deformations have been completely stabilized, and also after taking the necessary measures to stabilize and protect the surrounding areas.

Strengthening works and rehabilitation works are two different categories of interventions. While the first envisages the structural system, the latter concerns the thermal and comfort of the building.

The chosen strengthening solution is a decision usually made by the engineer. In this book, in Chapter 4, several strengthening solutions are given. Some of them are concrete/cement based, some of them are natural materials based. We know that some owners will prefer natural materials, and thus they can easily observe the contribution of the proposed and tested solution within the behavior of the wall, by looking at the envelope curves (produced by experimental testing on walls with and without strengthening) given in Tables 1 and 2.

The most common strengthening technique (using welded wire mesh and cement mortar) is further detailed in Chapter 5.4. This solution can be replaced with clay/earth mortar and another type of mesh (plastic, eco-friendly materials, etc.), the execution steps described being the same. But the engineers and owners choosing such solutions should be aware of the lower contribution of such a technique to the total resistance of the house in earthquakes, as shown in Tables 1 and 2. Of course, such solutions can be chosen in lower seismicity areas.

The TRAROM solution, involving diagonal timber planks, showed very good results and it is also based on timber, which is a natural material. In terms of workmanship, it is also less time-consuming and less skills required, so we recommend it as a replacement for the further described cement-based solution. However, we prefer to detail the latter, because its application is more complicated and our aim is that if used (and generally people prefer it), at least it should be correctly done.

In practice, observations have been made showing that the concrete materials applied on existing natural materials often increase the capillary action of water, harming the existing clay-based materials and in time enhancing biological decay. In this book we did not focus too much on the compatibility issue between materials, this not being our specialty, but only on the structural engineering aspect. Further research should be addressed on this topic, as a continuation of [21].

5.4.2 Proposed Interventions for Strengthening the Structural System

In order to strengthen the building presented in the previous chapter, it is necessary to carry out the following categories of interventions:

- Unloading the construction, by evacuating the objects stored in the attic, respectively the heavy filling from the slab structure and dismantling the ceiling structure.
- Demolition of the concrete access staircase from axis 1 and the perimeter sidewalk (which will be rebuilt later).
- Replacement of the degraded (rotten) portions of the timber parts of the set of structural walls with mixed composition (timber frames braced with clay brick masonry infill) and local repair of areas where the masonry is degraded.
- Jacketing the structural walls with mixed composition on both sides with reinforced plaster.
- Construction of a system of new foundations and plinths (executed so they are coupled and work together as an assembly) by underpinning the concrete in stages, reinforced with welded mesh.
- Demolition of the one-brick thick envelope walls and the support plinth in the alignment of axes A and between axes 4 and 5 (and of the access stairs), respectively 5 and between axes A and B, these being rebuilt from confined masonry (TFM) with a redundant composition (adopting a solution to increase the adherence between the masonry infill and the timber frame such as: inserting nails into the frame (Figure 37a), on the side which is in contact with the mortar, to be embedded in the mortar or using barbed wire, connected between the posts and embedded into the mortar joint (Figure 37b), respectively fixed to a new foundation system.
- Dismantling the old timber floor deck and replacing the damaged beam portions (if applicable) with new elements and restoring the connections to the walls.
- Rebuilding the floor deck with new antiseptic and fireproofed timber elements (two layers arranged overlapping and making 45° angles with the beam axes).

a b

FIGURE 37 Increasing adherence between the timber frame and infill: (a) nails inserted into the frame, before laying of the infill; (b) barbed wire, connected between the posts and embedded into the mortar joint.

- Strengthening the roof framework structure with clamps or metal plates (as the ones tested and described in Sub-Chapters 4.4.4 and 4.4.5).

5.4.3 Proposed Interventions for the Rehabilitation of the Strengthened Building

To rehabilitate the presented building, it is necessary to carry out the following categories of intervention:

- Restoration of the roof structure, consisting of new planks for the roof sheathing (antiseptic and fireproofed), anti-condensation foil, plank network, and metal tile panels cover.
- Replacement of the old joinery (windows and doors) with new energy-efficient elements, in the walls with a mixed structure, respectively the installation of new joinery at the level of the restored walls in the entrance hall area.

Note: For the walls with a mixed structure, the old door and window frames are preserved, these supporting the masonry infills located between the upper limit of the opening and the top plate sitting on the walls.

- Replacement of the thermal insulation between the slab beams with energy-efficient lightweight materials.
- Restoration of the ceiling structure.
- Restoration of the concrete access stairs.
- Installation of the thermal system on the outside of the envelope walls with methods proposed in the next chapter (Chapter 6).
- Installation of the soffit system under the roof eaves.
- Reconstruction of the perimeter sidewalk.
- Installation of a system of perimeter gutters on the roof contour, served by downpipes for the evacuation of rainwater.

5.4.4 Preliminary Works for Building's Strengthening

Quick glossary (what's what):

- Envelope walls → exterior loadbearing walls.
- Unprocessed timber strips (~7×5 cm) → old rough furring strips nailed under slab beams.
- Capillary gravel layer → granular layer of rocks with diameter more than 4 mm laid and compacted above ground that breaks rising damp.
- Plinths → lowest part of a building structure, typically a horizontal base or platform located between the ground level and the superstructure (walls, columns, or façade). It serves as the transition zone between the foundation and the vertical elements of the building and plays both structural and protective roles.
- Telescopic props → Adjustable vertical support used to carry floors temporarily; often placed on alignment beams and braced.
- Scaffolding/shoring – Temporary structures of props, frames, and braces used to support floors/walls and to provide safe access during works.
- Crossbeams/crosspiece → a horizontal support element placed at the top of a vertical prop (shoring post or adjustable steel support). Its main function is to distribute loads from above (such as beams, slabs, or formwork) over a wider area, ensuring safe and stable load transfer during construction or strengthening works.
- Benchmark (±0.00) → fixed elevation reference for the whole site.

Procedure step-by-step:

A) Roof works (safe access and partial unloading)
   - Remove the tile roof cover symmetrically, starting from the eaves up toward the ridge over a selected roof strip.
     Where: Along the roof contour above areas where slab beams bear on envelope walls.
     Why: Creates access/light and partially unloads the structure for safer work.
     Note: Above the entrance hall, remove tiles completely to fully unload the frame; the slab below is shored with scaffolding.

B) Initial floor/ceiling access & inspections
   - Lift 1–2 old floor boards near the supporting walls.
     Why: Quick inspection of beam bearings and wall connections.
   - Remove plaster from TFM walls on both faces, and rake out mortar joints 10–15 mm deep.
     Why: Exposes timber and masonry for inspection and future bonding.
   - Strip the ceiling in all rooms up to the rough timber strips (~7×5 cm) nailed to beam soffits.
   - Alternating removal of some rough strips (~7×5 cm)
     Where: First in the field between beams, then near the TFM walls, alternating to preserve stability.

Why: Opens inspection windows at beam supports without fully unbracing the system.
Stability note: Keeping some rough strips under beams together with the floor deck above preserves the global stability of the wall/ceiling assembly during later interventions (e.g., underpinning new foundations).

C) Remove infill and expose ground build-up
- Remove the heavy infill placed between slab beams, working between the remaining rough strips.
- Remove the floor decking system and the capillary gravel layer down to −0.215 m (top of the infill between plinths).
- Excavate the infill between plinths down to −0.565 m and evacuate the material.

D) Demolitions (external and internal elements)
- Demolish the unreinforced concrete access stair at axis 1 alignment and the degraded perimeter sidewalk.
- Demolish the heavy, slender chimney (controlled, from top down).

E) Local temporary support & targeted wall removal
- Locally support the end of the RC lintel (over openings on axis A, between 4–5) near the intersection of axes A and 4.
  How: Use crossbeams on a braced scaffold made of two pairs of telescopic props standing on timber studs, flanking the wall on axis A.
  Why: Provides a safe load path before adjacent demolition.
- Controlled demolition of the one-brick-thick masonry portion (full story height) tangent to the mixed-structure wall on axis 4, between axes A and B.
  Why: Opens access to intervene at the intersection of the TFM walls at the intersection between axis A and axis 4.

F) Openings and reference control
- Remove movable joinery parts from TFM walls, including side window sills.
  Why: Prevent damage and clear the work zone.
- Establish a permanent elevation reference (±0.00) at a safe, accessible location.
  Why: All levels (−0.215 m, −0.565 m, etc.) are measured from this datum.

Element map (where things sit):

- Tiles/eaves/ridge: top envelope; partial removal gives light/access and reduces roof load.
- Slab beams & rough strips: beams carry floor; rough strips brace the soffit—remove alternately.
- TFM walls: plaster removed both sides; joints raked for bonding and inspection.
- Infill & layers: deck + capillary gravel above; infill between plinths down to −0.565 m.
- RC lintel near the intersection between axes A and 4: propped by crossbeams + telescopic props before nearby demolition.

- Stair/sidewalk/chimney: demolished to clear degraded elements and unload.
- Benchmark (±0.00): fixed point for all level checks.

Tips:

- Keep symmetry when de-roofing; secure fall protection.
- Brace shoring; verify prop bearings on sound supports.
- During alternating strip removal, never remove all strips in a bay at once.
- Use dust control and manual tools near TFM timber to avoid accidental damage.
- Tag and store removed joinery if to be reinstated.

5.4.5 Works for Replacing Degraded Wooden Parts and Masonry in TFM Structure Walls

Following the structural investigations carried out, degraded areas were observed in the TFM structure walls both at their lower part and at the plinth level, an aspect that certifies the presence of infiltrations. The confined humidity led, over time, to the rotting of some parts of the wooden elements and the deterioration of some masonry elements, as well as the plain concrete plinth. At the same time, due to the displacements of the plinth (locally significantly degraded) produced by the variations in the volume of water in the soil during repeated freeze–thaw cycles (the plinth was placed directly on the ground), in some locations the masonry at the lower part of the walls presents dislocations.

The works presented further in this sub-chapter will be carried out, depending on the situation, at the level of the TFM walls that present degradations.

5.4.5.1 Stabilization of the Existing Masonry Infills Located Above the Degraded Areas

The idea is to create a temporary—but strong—reinforced-plaster jacket on the upper part of the TFM walls, so the wall stays stable while you remove and rebuild the degraded bottom portions (mudsills, posts/bracings ends, and infill). A final result is shown in Figure 38a to Figure 38d, while the steps explained thoroughly are shown in Figure 39a to Figure 39f.

Quick glossary (what's what):

- Assize/course = one horizontal brick row/layer.
- Mudsill = The lowest horizontal timber of a frame wall, resting on the plinth or foundation.
- Welded mesh panels = steel reinforcement mesh (to be plaster-jacketed).
- Reinforcement coupons = short rebar pieces inserted through drilled holes at a horizontal joint.
- Clips = small steel wire ties/clamps anchored into drilled holes to hold the mesh off the wall.
- Rough/spatterdash coat = thin, gritty bonding coat applied before plastering.

- Plaster "thickness guides" = small vertical strips of mortar set first, used to rule the final coat flat.

Procedure step-by-step:

A) Investigate and set the control level
   - Survey the TFM wall:
     What: Identify rotten timber (posts/bracings ends) and disintegrated infill; mark all degraded zones on the wall.
   - Choose a horizontal joint/course just above the highest degraded point (measured from wall base). Mark this joint.
     Why: This will be the control level where we pass reinforcement through the wall to support the upper mesh/plaster while the bottom is opened.
   - Expose sound timber sections at the damaged ends of posts and bracings (open locally to see intact wood).
     Site note: Highest degraded masonry was found at ~7–8 courses (~60 cm) above floor, where sound timber can still be identified.

B) Plan mesh and drill the horizontal support line
   - Size and count welded-mesh panels for the upper wall zone, allowing ≥3-mesh overlaps; measure from the marked joint up to the top of wall.
   - Mark hole positions along the marked horizontal joint, only above degraded areas, spaced ~20–30 cm.
   - Drill the holes/channels along that joint.
   - Insert rebar coupons through the holes so they stick out ~10 cm on both faces.
     Why: These coupons pass through the lowest mesh row of the upper panels and carry/support the wall after you remove the degraded bottom infill.

C) Make plaster thickness guides & leave overlap room
   - Mark plaster thickness points in a grid: ≈1.50 m horizontally and ≈0.50 m vertically, up to ~1.00 m above the wall base. Check all marks are coplanar with a straightedge.
   - Form vertical mortar "guides" between those marks (after wetting the wall), creating parallel vertical strips that set the final plaster thickness.
     Important: Do not cover the lower ~3 mesh rows zone (keep it free) — this is where upper meshes will overlap later with lower meshes installed after repairs at the base.

D) Pre-wet and apply the upper reinforced plaster (leave bottom ~1 m free)
   - Wet the surfaces to be plastered (both faces).
   - Apply plaster evenly from the top downward to ~1.00 m above the base, on both wall faces.
     How: The plaster will flow behind the welded meshes (once placed) and embed them; rule the surface top-to-bottom using the vertical guides.

   Why this gap?
   - The lower ~1 m remains unplastered (for now) so the upper meshes can overlap ~3 meshes with the new lower meshes that will be added

after you fix the degraded base and cast the new foundations + plinth (underpinning).
- The upper reinforced plaster already locks the upper masonry and keeps it stable while the bottom is opened.

E) Fix mesh to the wall with clips; prepare the bond coat
- Lay out mesh panel positions on both faces (allow 3-mesh overlaps in both directions) and mark clip locations along the panel lines (per the design).
- Drill holes for clips (steel wire/clamps) along the mesh runs, taking care to avoid vibration that could crack the existing infill.
- Insert the clips into the drilled holes and set them with mortar (cement or clay/earth), after properly wetting the surrounding masonry.
- Cover clip ends (sticking out) with adhesive paper tape to keep them clean.
- Wet the masonry above the damaged areas and apply a uniform spatterdash (rough/bond) coat with additives up to the marked horizontal joint.
- Remove the tape from clip ends.

F) Install and fix the upper mesh panels
- Place welded mesh panels (to project sizes) on the spatter-dashed surface between the top of wall and the marked joint, supported on the clip ends.
  - At wall intersections, use L-bent panels (90°) and overlap ≥3 meshes (Figure 38c).

FIGURE 38 Stabilization of the existing masonry infills located above the degraded areas: (a) protruding starter bars are passed from one side to the other, to later make the connection between the wall's jacketing and the foundation; (b) at the intersection of the walls, horizontal bars are connected to the perpendicular wall; (c) welded mesh panels bent to a 90° angle are laid and overlap by 3 meshes; (d) outside, the protruding starter bars are bended at their ends, waiting to be connected to the vertical starter bars coming from the underpinned (new) foundation.

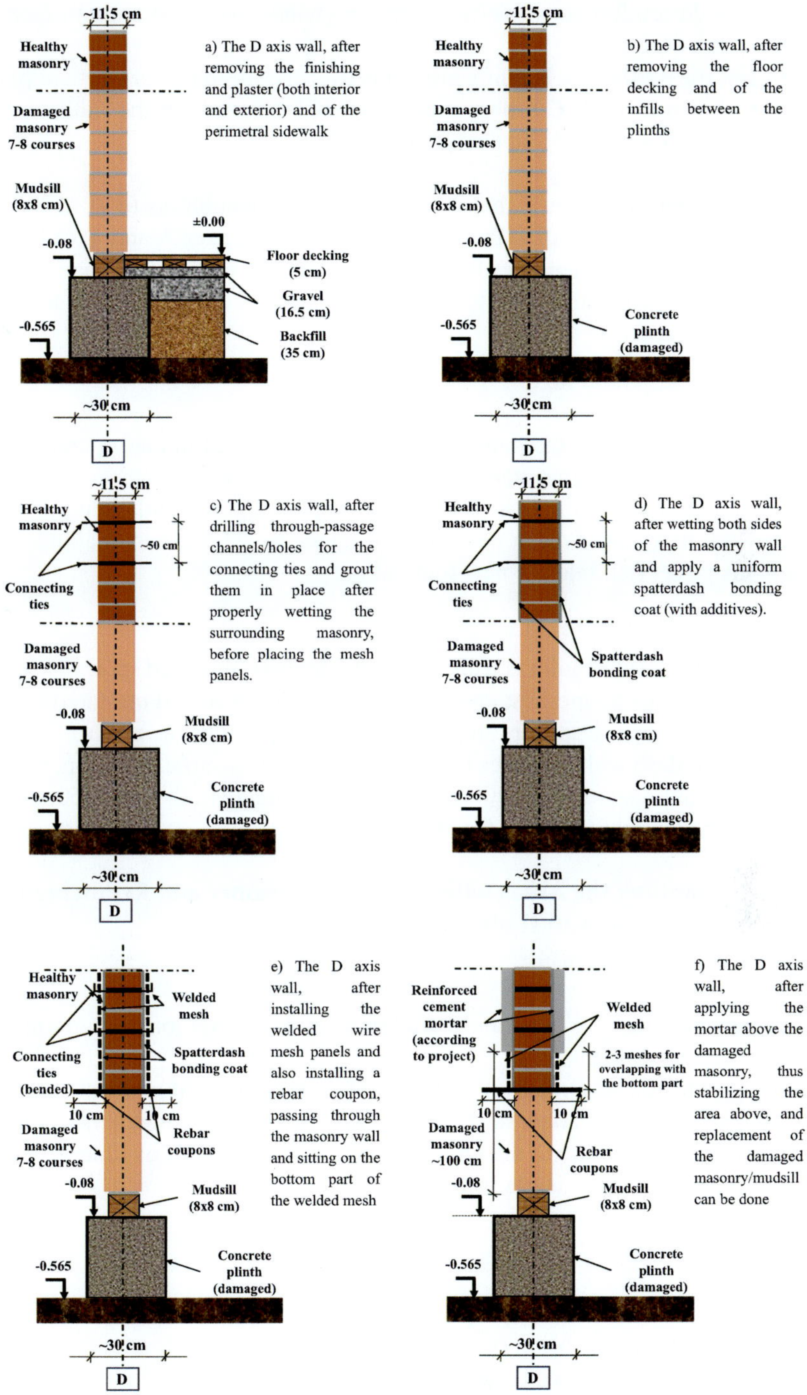

FIGURE 39 Steps to strengthen the TFM walls, at the upper part, for their stabilization while replacing the damaged bottom masonry and rotted timber, using welded mesh.

- Add reinforcement above openings (windows/doors) on both faces per the design.
- Fix the meshes by bending clip ends over the mesh nodes (vertical plane).
  - Insert local PVC/timber spacers to hold the mesh off the wall so plaster can flow behind and embed it.

Tie-column interface note: If this wall meets a RC tie-columns (e.g., at the edge of the one-brick envelope wall at the intersection between axes A and 4, or between axes A–B and axes 4–5), then after fixing meshes on both faces, install pairs of horizontal starter bars (projecting bars) to connect the jacketed TFM wall edge to the RC tie-columns (per project).

Why this works (structural logic):

- The rebar coupons at the joint + upper mesh embedded in plaster create a stiff diaphragm-like skin that stabilizes the upper masonry.
- Leaving the lower band free allows you to open, replace, and underpin safely; later, the new lower mesh overlaps the upper by ≥3 meshes, making a continuous jacket tied into the new foundation.

Tips:

- Recheck the control joint is truly above the highest degraded spot.
- Keep mesh laps ≥3 meshes; ensure clips are well seated and mortared.
- Maintain cover: use spacers so plaster fully encapsulates the mesh.
- Use spatterdash with additives for good bond; pre-wet surfaces (damp, not wet).
- Record hole spacing (20–30 cm), guide spacing (≈1.50 m × 0.50 m), and unplastered height (~1.0 m) for checks.

5.4.5.2 Dismantling the Areas with Degraded Masonry and Removing of Rotten Portions of Timber Elements

Quick glossary (what's what):

- Alignment beam → a straight timber/steel beam set on the ground that distributes prop loads.
- Soffit – underside surface of a structural or architectural element, such as a beam, slab, arch, balcony, staircase, or eave. In building construction, the term most commonly refers to the underside of a floor slab, roof overhang, or structural member that is visible from below.

Procedure step by step:

A) Set alignment beams and props
  - Mark alignment-beam positions at –0.565 m (ground level reference).
    Where: Along zones adjacent to walls where you'll intervene.
    Why: These beams will carry the props safely.
  - Place the alignment beams on the marks.

What: Seat them firmly on compacted ground; add sole plates if needed for bearing.

- Build the shoring system for the targeted slab areas.
  How: Set props on the alignment beams; at the top, place crosspieces perpendicular to the slab beams so the slab beams bear on the crosspieces.
- Pre-load (lightly jack) the props.
  What: Raise the prop heads slightly so the shoring is snug between the –0.565 m base and the slab-beam soffit near the walls.
  Why: Provides a safe load path before you touch the wall base.

B) Open up the degraded wall base (controlled)

- Carefully dismantle the degraded masonry at the wall base.
  Where: Lower part of the walls, down to the lower edge of the already-installed reinforcement meshes (the level with the horizontal bars/coupons).
  Why: Exposes the rotted ends of posts and braces and the degraded mudsill above the plinth while the upper wall is stabilized by the mesh.

C) Prepare and remove degraded timber

- Mark the cut lines on the lower ends of posts, diagonals/braces, and on the degraded mudsill segments.
  How: Mark beyond visible decay into sound wood.
- Cut along the markings and remove the degraded portions.
  Sequence: Posts/braces ends and mudsill segments are taken out only after shoring is verified and the upper mesh jacket is in place.

Element map (where things sit, in section)

- Ground (–0.565 m) → alignment beam sits here; props stand on it.
- Props → crosspieces → slab-beam soffit → vertical load path upwards to the floor structure.
- Wall base → plinth at bottom; above it the mudsill; then TFM (posts/braces + infill).
- Degraded zone → lowest masonry + rotted timber ends to be removed.
- Reinforcement meshes above → already fixed; their lower edge is the upper limit of temporary demolition.

Tips:

- Do not over-jack: just snug the shoring; recheck after any settlement.
- Use bracing between props; ensure crosspieces are tied off and centered under slab beams.
- Protect/avoid damaging the reinforcement meshes at their lower edge (overlap zone for later works).
- Remove masonry progressively; avoid prying that could vibrate the TFM frame.
- Confirm utilities are absent; wear PPE; maintain clear egress paths.
- Document cut lengths and the extent of removal for the repair logs.

5.4.5.3 Temporary Restoration of the Flatness of the Horizontal Surface of the Unreinforced Concrete Plinth in the Degraded Areas, In Order to Restore the Damaged Portions of the Timber Mudsills

The existing plinth is partly missing/uneven/spalled and will later be replaced by a new plinth tied to the new foundation (after wall repairs). Until then, we need a continuous, flat bearing surface to safely splice/replace timbers and rebuild masonry. We achieve this with temporary low-grade concrete infill (easy to break out later) and leveling mortar (Figure 40g).

Quick glossary (what's what):

- Low-grade concrete (temporary): weak mix used to temporarily fill gaps, easy to remove later.
- Bushing/bush hammering – a surface treatment technique used mainly on concrete, stone, or masonry surfaces to create a rough, textured finish. It involves impacting the surface with a bush hammer — a specialized tool with a head covered in small pyramidal or pointed teeth — either manually or mechanically.
- Formwork: temporary molds (timber/plywood) that contain fresh concrete to the desired shape until it sets.

Steps:

- Identify degraded zones (deep and surface) of the plinth.
  Where: Along the wall base; note missing, cracked, spalled areas.
  Why: Defines where temporary infill/leveling is needed.

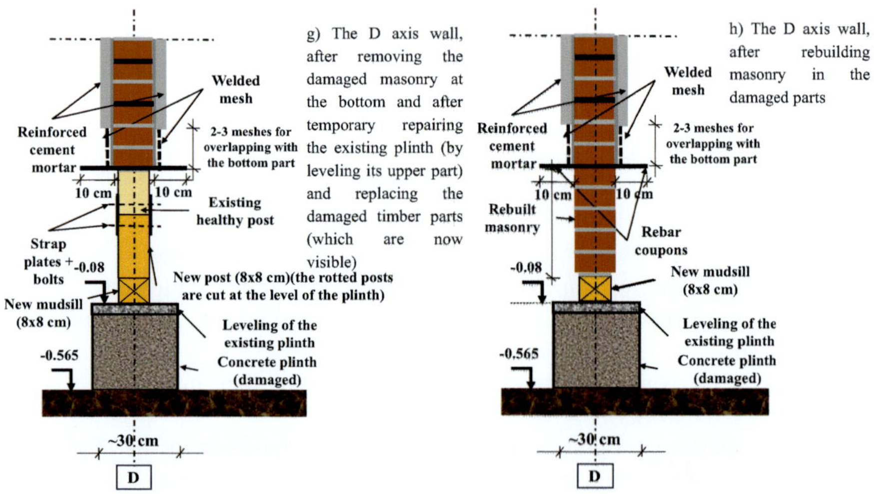

FIGURE 40 Steps to replace the rotten timber parts (post and mudsill) in the bottom part of the walls.

- Clean recessed/void areas thoroughly.
  What: Remove dust, debris, loose mortar/concrete, contaminants.
  Why: Ensures bond and avoids trapping friable material.
- Install timber side formwork along deteriorated runs of the plinth, with the top edge set at –0.08 m.
  Why: Controls the finished level and straightness of the temporary layer.
- Pre-wet the old concrete (SSD: saturated surface-dry).
- Place low-grade concrete in the missing portions, compact (rod/trowel), and strike off to the formwork top.
  Why: Temporarily reconstructs continuity where the plinth is missing; remains easy to remove when the new plinth/foundation is built.
- In uneven (but not missing) areas, pre-wet and apply leveling mortar, trowel to the formwork top.
  Why: Achieves flatness and coplanarity across the wall length for accurate timber/masonry reinstatement.

Element map (where things sit, in section)

- Existing RC plinth (base) → locally cleaned/bush-hammered.
- Temporary low-grade concrete → fills gaps/voids to the level –0.08 m.
- Leveling mortar → skins high/uneven areas to the same level.
- Formwork → side boards that set the line and level during placement.

Tips:

- Bush-hammer to sound concrete; remove all loose material.
- Pre-wet to SSD; avoid standing water before placing.
- Keep the –0.08 m level consistent; check with a straightedge and level.
- Don't rely on temporary concrete for structural capacity; it's a working surface only.
- Protect from rapid drying (light curing) so surfaces don't dust or craze before use.
- Record lengths/locations of temporary infills for planned break-out when the new plinth and foundation are executed.

5.4.5.4 Replacing the Damaged Timber Members of the Wall Frames

This step is done right after strengthening the upper part of the masonry in the TFM walls (even if not damaged) and also after removing the damaged masonry at the bottom part of the walls, thus the timber elements are basically exposed, as shown in (Figure 40g). Then, from this point, the procedure is done as described in this sub-chapter. In this sub-chapter, additional to the 5.4.5.2 sub-chapter), other timber elements are mentioned, not only the ones in the exposed bottom parts of the stabilized walls.

Quick glossary (what's what)

- Post = vertical timber in the frame.
- Mudsill (sill) = horizontal timber on top of the plinth.

- Diagonal/brace = inclined timber element that stiffens the frame.
- Strap plate + bolts = steel side plate(s) bolted to timber to form a splice.
- Staples/clamps = temporary side clamps to hold parts before final fixing.
- Plinth top = reference level –0.08 m.

Step-by-step procedure:

- Measure each replacement piece individually.
  What: Determine the clear splice length for every member whose deteriorated end was removed.
  Why: Length depends on member role (post/brace/sill) and how it intersects with adjacent timbers.
- Cut new dressed (planned) timber to size (8×8 cm).
  What: Cut to measured length and shape the ends (square, bevel, or notched) so they fit the existing assembly.
  Why: Proper end geometry ensures full contact and good load transfer.
- Lower post at wall intersections (where applicable).
  Place: Fit the new post extension at the bottom end of the corner/intersection post.
  Fix:
  - Upper splice: Steel strap plates + bolts to the sound part of the post.
  - Lower temporary hold: Staples/clamps to the horizontal sill until final fastening to the new plinth.
    Important: The added piece must NOT go below –0.08 m (top of plinth), because the final anchorage will be with steel fasteners embedded in the new plinth.
- Close the mudsill (sill) run.
  Place: Install the new sill segment where the old was removed.
  Fix: Strap plates + bolts at both ends to the existing sill pieces.
- Lower ends of intermediate posts.
  Place: Install the new post foot piece.
  Fix: Strap plates + bolts to (a) the post above and (b) the mudsill on the plinth.
- Lower ends of diagonals/braces.
  Place: Install the new brace foot piece.
  Fix: Strap plates + bolts to (a) the brace above and (b) the mudsill on the plinth.

Element map (where things sit):

- Plinth (–0.08 m top): carries the mudsill; posts and braces bear into the sill.
- Post/brace splices: strap plates on both sides across the joint; bolts pass through timber + plates.
- Temporary clamps: used only to hold a corner post extension until the new plinth anchors are in.

Tips:

- Splice length: extend into sound wood; follow the project for minimum lengths.
- Bolt detailing: pre-drill; keep edge distances and bolt spacing per design; use washers.
- Timber condition: use dried, dressed, antiseptic + fire-retardant treated timber; seal cut faces.
- Bearing faces: ensure full, flush contact; packers/shims are not a substitute for proper cutting.
- Alignment: check plumb (posts) and line (sills/diagonals) before final tightening.
- Final anchorage: do not rely on temporary clamps; posts at intersections get steel fasteners into the new plinth when it's cast.

5.4.5.5 Works for Rebuilding Masonry in the Areas Where the Bond Was Dismantled Within the Wall Structure

Rebuilding of masonry in the zones dismantled in a controlled manner shall be carried out uniformly (Figure 40h), up to the level of the joint where stabilization parts (with welded mesh panel, as described in 5.4.5.1 sub-chapter) were fixed and around the intersections of intermediate posts and diagonals with the mudsill. At the immediate vicinity of the posts at wall intersections (to be connected to the new RC plinth by steel fittings embedded in concrete), leave an unbuilt gap with a length of about one brick and a height up to the limit of the already fixed meshes. Procedure step-by-step:

- Wet the surfaces of the old masonry along the contour of the zones where controlled demolition was performed.
- Lay bricks (of the same type as those used in the structure) in the respective locations, after proper wetting and surface drying, using masonry mortar with a recipe in accordance with the strengthening project. During laying, wedge bricks with mortar between the timber members of the wall structure, adjusting brick dimensions as required (breaking them with an adze or small hammer). Ensure proper bonding and fully filled joints. Before laying the bricks, for an increased adherence between the timber frame and infill, 10 cm nails can be inserted into the timber frame on the side which will be in contact with the mortar, thus in turn the nails will be finally embedded in the mortar too.
- After wedging with mortar the last units, cut the ends of the rebar dowels that pass through the lower marginal openings of the installed meshes.

5.4.6 Execution of the Foundation System and New Plinth under TFM Walls Using Underpinning Technology

Although in some cases a stone or clay brick foundation may be also executed, from an engineer's point of view in a seismic country, the best solution for a solid foundation is RC. Therefore, such a solution is further described.

Quick glossary (what's what):

- Good bearing soil (below frost depth) — Competent soil strata located deeper than the local freezing line to avoid frost heave and seasonal movement.
- Frost depth — Maximum depth to which ground water in soil freezes; foundations must bear below this level.
- Capillary water rise — Upward movement of moisture through pores in soil/ concrete; controlled by waterproof concrete and detailing.
- Underpinning "in plots" (60–80 cm) — Phased excavation and casting of short, alternating segments to avoid destabilizing the house.

Procedure described:
To ensure proper load transfer from walls to the bearing soil, avoid exposing the structural system to frost-depth soil movements, and prevent capillary water rise toward the base of TFM walls (with embedded timber), an efficient foundation system of waterproof concrete must be executed, with the footing bearing on good soil below the local frost depth, and a new RC plinth anchoring the starter bars that continue the vertical bars of the wall jacketing meshes.

This continuous foundation system, coupled with the plinth under the building walls, shall be executed by underpinning in plots of 60–80 cm long (Figure 41a) (to avoid the destabilization of the house), with a foundation block width of 50 cm. Depth should be according to the strengthening project and geotechnical study; concreting top elevation being at (–0.60 m).

FIGURE 41 Underpinning the foundation: (a) Making the excavation in plots; (b) the plinth above the excavation is made with gravel and sand (very weak); (c) the excavation is done at the corner in both directions.

The RC plinth shall be executed after the foundation block concrete has set; segments have the same length, width 30 cm, height 55 cm; the formwork's top edge reaching the elevation –0.05 m. Thus, the concrete level in the new plinth exceeds by 3 cm the placement elevation of the timber mudsills (–0.08 m), ensuring, after setting, perfect contact between TFM walls and the plinth executed segmentally beneath them.

Trenches parallel to the walls and the excavations for underpinning plots shall be executed in stages (with a set number of plots per stage), in the sequence specified in the strengthening project, following the next rules:

- The weakened area of the old plinth footing (through which the walls discharge to soil until underpinning is complete) shall not exceed 25% of the total area.
- Between two excavated segments, leave 2–3 unexcavated segments.
- First excavated and concreted segments shall be at the worst defects of the old plinth.
- Excavations start from the ends and close toward the middle.
- At corners and intersections of the old plinth, excavate and cast a segment that includes all intersecting wings (Figure 41c).

For envelope (exterior) walls, trenches for underpinning plots under the existing plinth shall be outside; for interior walls, place trenches on one side, depending on room size and shoring layout.

5.4.6.1 Setting Out of Underpinning Plots and Erection of Shoring to Support the Floor (Stabilizing Role)

Quick glossary (what's what):

- Alignment beams (at –0.565 m) — Straight timbers/steel placed on the ground at the set level to distribute loads from props and define shoring lines.
- Slight jacking of prop heads — Gently tightening/raising props so the shoring fits snugly ("wedged") between base level and the slab beam soffits without overloading.

Procedure step-by-step:

- Set out the widths of trenches parallel to the wall plinths and close to them at elevation –0.565 m, from which soil will be excavated for each underpinning plot (Figure 42j).
- Mark the limits of each plot on the plinths at the lengths provided in the project and number them.

Note: At plinth intersections, along one direction the plot limit shall coincide with the vertical face of the plinth in the orthogonal direction.

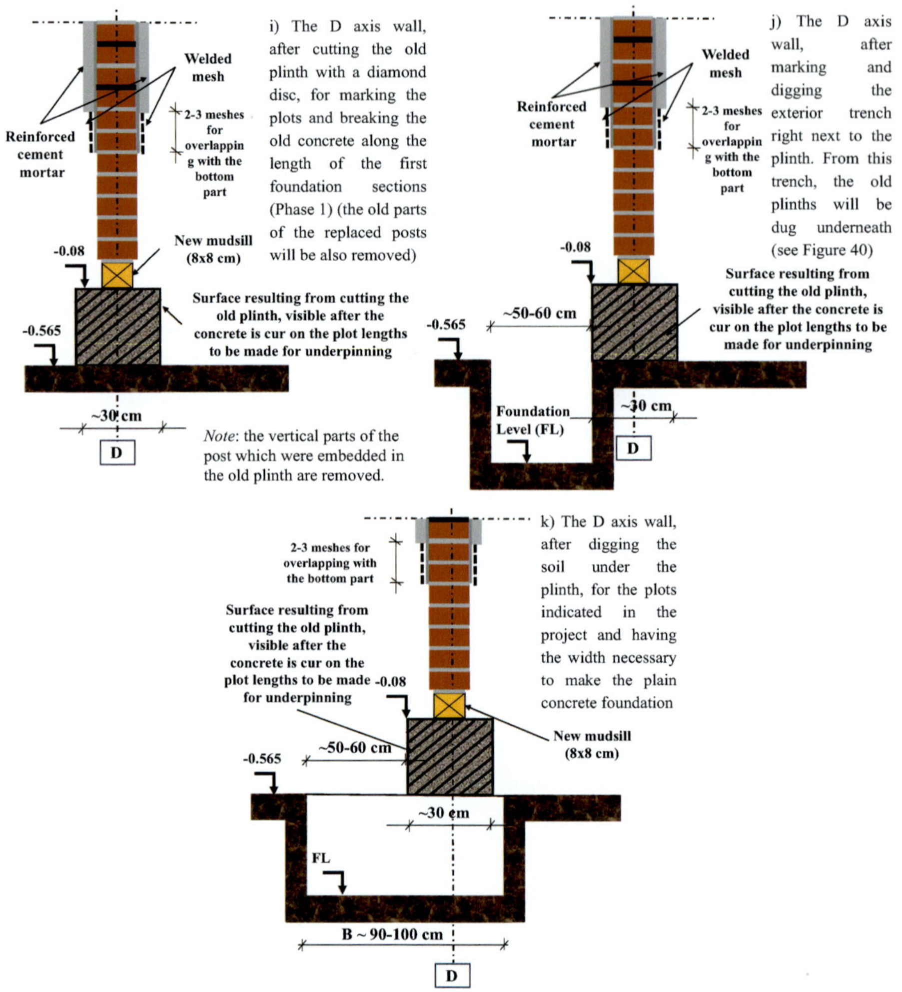

FIGURE 42 Execution of the new plinth's plots using underpinning technology.

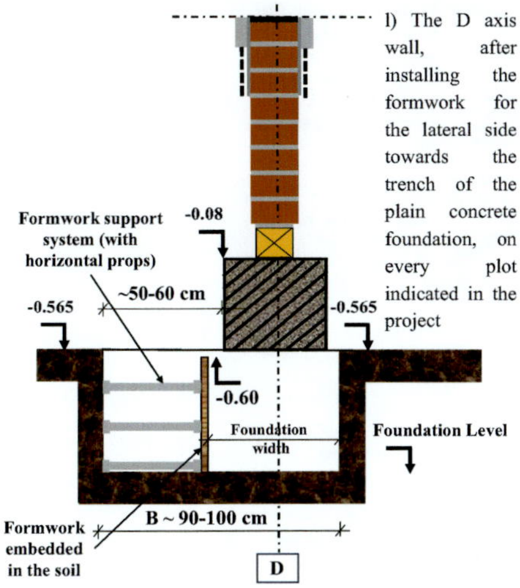

l) The D axis wall, after installing the formwork for the lateral side towards the trench of the plain concrete foundation, on every plot indicated in the project

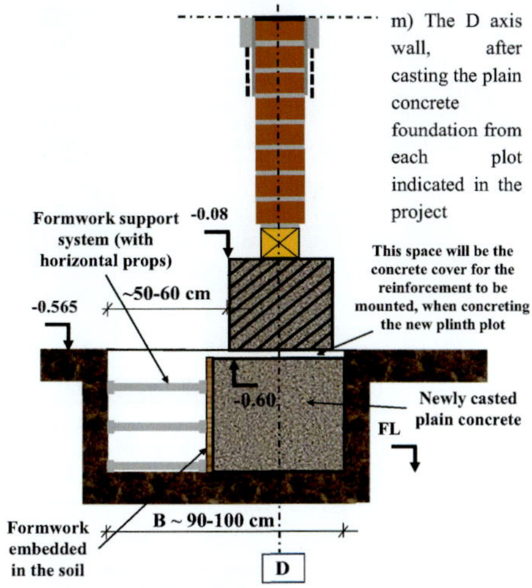

m) The D axis wall, after casting the plain concrete foundation from each plot indicated in the project

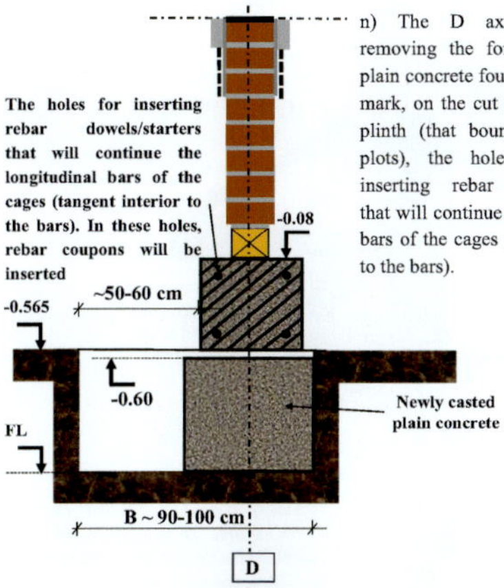

n) The D axis wall, after removing the formwork on the plain concrete foundation plot and mark, on the cut faces of the old plinth (that bound the Stage 1 plots), the hole positions for inserting rebar dowels/starters that will continue the longitudinal bars of the cages (tangent interior to the bars).

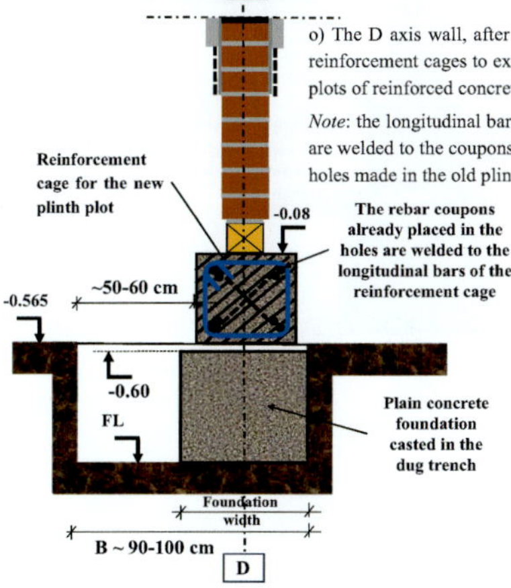

o) The D axis wall, after installing the reinforcement cages to execute the new plots of reinforced concrete plinth

Note: the longitudinal bars of the cages are welded to the coupons placed in the holes made in the old plinth

FIGURE 42 (Continued)

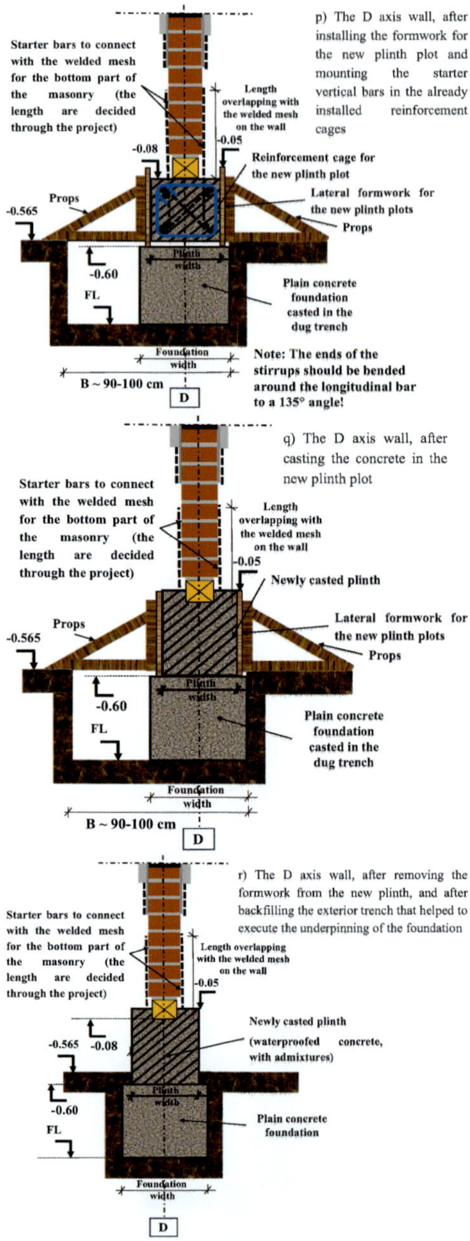

FIGURE 42 (Continued)

- Set out the positions of alignment beams for the shoring that supports the floor (also stabilizing the wall system during excavation) at elevation –0.565 m, ensuring safety distances from trench edges.
- Place the alignment beams on the markings.

- Erect the shoring (properly braced); props bear on alignment beams; slab beams bear on props via crosspieces perpendicular to the beam's direction.
- Slightly jack the prop heads until the shoring is wedged between base level and the soffit of the slab beams.

5.4.6.2 Dismantling Masonry at Wall Intersections (Where No Repairs Were Needed) to Enable Fixing the Post Bases to the New Plinth

Quick glossary (what's what):

- Steel bearing fasteners (at posts) — Metal base/anchor components that will seat/anchor timber posts into the new RC plinth.
- Flat end sections — Clean, even concrete faces created at plot boundaries to aid formwork, dowelling, and bonding.

Procedure step-by-step:

- Mark at the bases of intersections of TFM walls (without prior repair works) the zones of controlled dismantling of the masonry bond near the timber posts, to install steel bearing fasteners in the new RC plinth.
- Carefully dismantle intact masonry at the wall bottom parts near intersections over a length of about one brick (measured from the post face) and a height of about 7–8 courses (~60 cm).
- Temporarily fix the intersection posts to the mudsills with side-driven staples/clamps.
- Cut the old plinth's concrete along the Stage 1 plot layout (using a diamond saw) (Figure 42i).
- Break out the concrete between cuts (plot limits) to obtain flat end sections. Remove debris and decayed timber remnants from some intersection posts.
- Cut the intersection posts at the lower limit of the adjacent mudsills (–0.08 m) after breaking out the surrounding old concrete.

5.4.6.3 Works for Executing the Foundation Plots in Stage 1

Quick glossary (what's what):

- Design foundation level — The target founding elevation from the strengthening project (drawings/specs) where the new footing bears on competent soil.
- Groundwater lowering (dewatering) — Pumping or wellpoint systems used to keep the excavation base dry and stable if groundwater is high.
- Uncharted utilities — Service lines not shown on plans; if suspected/found, excavation stops until the owner/operator inspects and authorizes actions.
- Vertical side formwork (TEGO plywood) — Smooth-faced plywood panels used to close the open side of each plot excavation; top set to –0.60 m, braced to the trench wall (Figure 42k).

- Blinding concrete layer — A thin, flat concrete bed placed on the excavation base to provide a clean working surface for rebar and formwork.

Procedure step-by-step:

- Successively excavate soil from the trenches adjacent to the Stage 1 plots, over manageable lengths (manual excavation), down to the design (according to the strengthening project) foundation level of the underpinning plot footings; stockpile the spoil (Figure 42j).
- Mark the vertical limits of the Stage 1 plots on the trench walls (continuations of the cuts made in the plinth) down to the base level.
- Successively excavate soil (starting from trench sides between the marked vertical limits) from the volumes of Stage 1 plots under the walls, achieving the designed geometry and dimensions; stockpile the spoil.

Note**:** During excavation:

- Depending on depth, soil characteristics, and groundwater level, apply appropriate and recognized methods to prevent trench wall collapse.
- Lower the groundwater below the excavation base if required (system selection based on the geotechnical study).
- If uncharted utilities are suspected during works, stop excavation and notify the client/operator; decommission only with their consent and supervision.
- The contractor shall call the geotechnical specialist on site at foundation level and whenever discrepancies with the study are observed.
- If cracks appear at the excavation base, the designer shall set the necessary measures.
- Do not continue any construction work before obtaining the designer's approval of the excavation level and founding soil characteristics.
- Transport excavated soil by dump truck to designated stockpiles.
- Install the vertical side formwork (TEGO plywood) closing the fourth side of the Stage 1 plot excavations, the top edge being at –0.60 m; clear distance to the opposite trench wall equals the foundation width (based on the strengthening project). Brace the formwork against the trench wall (Figure 42k).

Note: Where sewer/water piping crosses the continuous underpinning foundation, widen the trench locally beyond the foundation width. Cast a blinding concrete layer; place rebar cages (based on the strengthening project) on spacers for cover. At pipe crossings, interrupt the face reinforcement (ends bent at 90°, with circular collars and pairs of diametrically opposed bars inclined at 45°); install capped PVC sleeves equal to foundation width, with internal diameter greater than the pipe's outer diameter. Longitudinally, close the underpinning plot with two parallel braced formwork faces (between which the PVC sleeves are fixed), with top edge at (–0.60 m). Provide spacers at cages for cover at formwork faces.

- Successively pour concrete in the Stage 1 underpinning plots and compact adequately; strike off to elevation (–0.60 m) (Figure 42l).

Note: Use waterproofing admixtures in the concrete for underpinning plots.

- After the concrete reaches sufficient maturity, strip the Stage 1 plot formwork and take measures to preserve moisture in the concrete mass.

5.4.6.4 Fabrication of Cage Segments and Positioning for the Plinth Segments in Stage 1 Plots

Quick glossary (what's what):

- Reinforcement cages — Prefabricated steel bar assemblies made of longitudinal bars and stirrups, forming the reinforcement skeleton of each new plinth segment.
- Rebar dowels/starters — Short steel bars inserted and driven into drilled holes in the old concrete to connect new reinforcement to existing structure.

Procedure step-by-step:

- Fabricate reinforcement cages for the new plinth segments included in Stage 1, according to the design project of the strengthening works and make measurements between end sections obtained from cutting the plinth.
- Mark the position of the new plinth formwork on the (–0.60 m) surface of the Stage 1 foundation plots.
- Position and center the plinth cages on the (–0.60 m) surfaces of the cast plots, supported on planks.
- Mark, on the cut faces of the old plinth (that bound the Stage 1 plots), the hole positions for inserting rebar dowels/starters that will continue the longitudinal bars of the cages (tangent interior to the bars) (Figure 42m).
- Remove the cages and drill the holes to the depths specified.

5.4.6.5 Final Assembly of Plinth Cage Segments and Steel Fastening for Fixing Intersection Post Bases on the New Plinths (Stage 1 Plots)

Quick glossary (what's what):

- Stirrups bundled toward the ends — Temporarily sliding groups of transverse ties to the cage ends to clear space for welding/fastening; later redistributed.
- Intersection post — Vertical timber member at a wall junction (corner/tee) bearing on the plinth.
- PVC foil on old plinth cuts — Plastic sheet applied to existing concrete cut faces to control adhesion/contamination at staged joints.
- Spacers (for concrete cover) — Small blocks/clips that keep rebar at the correct distance from formwork, ensuring required side/bottom cover thickness.

Procedure step-by-step:

- Reposition reinforcement cages on the –0.60 m surfaces of the cast Stage 1 plots at wall intersections; shift the stirrups in bundles toward the ends.

- Bring the steel fasteners to the base of the intersection posts and position it according to the project, ensuring its underside bears on the concrete at –0.60 m.

Note: To position the fasteners, depending on design, the length of the mudsill (in contact with the intersection post) may be slightly shortened under the wall perpendicular to the one whose plinth was cut along the plot.

- Mark fixing-holes positions on the post's surface through the fastener's holes.
- Remove the fastener, then drill the holes in the posts.
- Reposition the fastener at the post bases and fix with bolts inserted through the drilled holes.
- Move the stirrup bundles toward the intersection posts.
- Place PVC foil on the concrete faces of the old plinth cuts; insert rebar dowels/starters into the drilled holes and hammer them to the required embedment.
- Weld the rebar dowels to the longitudinal bars of the cages.
- Redistribute the stirrups along the cages and tie at the marks; fix spacers to allow a proper thickness for the concrete cover (sides and bottom).
- Wet the old masonry surfaces around the zones where the bond will be rebuilt near the post bases.
- Lay bricks (same type as existing), after proper wetting and drying, using design project-specified mortar; wedge bricks with mortar between the timber members, adjusting dimensions accordingly.

5.4.6.6 Final Assembly of Plinth Cage Segments in the Central Part of the Walls (Stage 1 Plots)

Quick glossary (what's what):

- Wall field zones — The wall spans between structural axes (between columns or tie elements) where continuous plinth reinforcement is being built.
- End stirrups — The closed transverse bars at the ends of the cage; temporarily moved inward ("bundled") to clear the ends of the longitudinal bars for welding.

Procedure step-by-step:

- Reposition reinforcement cages on the –0.60 m surfaces of the cast Stage 1 plots in the wall field zones.
- Shift end stirrups into a central bundle, freeing the longitudinal bar ends.
- Place PVC foil on the concrete faces of the old plinth cuts; insert and hammer rebar dowels into the drilled holes to the required embedment.
- Weld dowels to the cage longitudinal bars (Figure 42n).
- Redistribute and tie the stirrups at marks; fix spacers for concrete cover (sides and bottom).

5.4.6.7 Concreting the New Plinth Segments (Stage 1 Plots)

Quick glossary (what's what):

- Lap length — The required overlap between two rebars so forces can transfer safely (per design/code).
- Waterproofing admixtures — Additives mixed into concrete to reduce permeability and improve resistance to water ingress.
- Hardened laitance on protruding bars — Brittle, cement-rich film that can form on exposed rebars; removed after ≥24 h to ensure good bond for later work.

Procedure step-by-step:

- Mark the positions of starter bars that will anchor the vertical bars of the wall meshes at each wall base.
- Insert the starter bars from top to bottom into the cages of the new plinth segments in Stage 1, place at the marks ensuring lap length, then tie with wire (Figure 42o).
- Install side formwork for the plinth segments; top edge at –0.05 m (Figure 42o).
- Wet the timber formwork about 2 hours before casting; remove any water pooling at the base.
- Pour and compact the concrete in the new plinth segments, preventing air pockets under the wall mudsills (Figure 42p).

Note: Use waterproofing admixtures in the plinth concrete.

- Strike off the concrete with a batten guided by the formwork top edge.
- After at least 24 hours, remove the hardened laitance from the protruding starter bars on the plinth surface.
- When the concrete reaches the required maturity, strip the side formwork and treat exposed concrete surfaces accordingly (Figure 42, q).

General note for subsequent underpinning/plinth stages:

- Clean exposed plot faces that were in contact with soil by scraping and wire-brushing to remove all soil traces.
- When exposing end faces of freshly cast plinth segments (by breaking the old plinth), remove the PVC foil, bush-hammer the faces, and clean the now-visible rebar dowels.
- As cages are installed in successive plinth segments, the longitudinal bar ends are welded to dowels protruding from previously cast plots to achieve continuous reinforcement beneath each wall. Also the starter bars are marked and installed to anchor the vertical mesh bars of the wall jacketing.
- The time between concreting one segment and excavating the adjacent segment shall be at least 24 hours.

- Once the plots along an exterior wall are connected and the rigid foundation completed in that location, execute compacted backfill (in layers) to close the trench.

5.4.7 Earth Backfilling for Trenches Used in Underpinning Works

Quick glossary (what's what):

- Sleeve caps — Temporary end caps on sleeves to keep debris out; removed before installing pipes.
- Sewer pipes with direction-change fittings — Wastewater piping assembled piece by piece using bends/tees to route from sleeves to the designated exit above the ground-floor slab.
- PPR water pipe — Polypropylene random copolymer pressure pipe for cold/hot water distribution; routed to its planned location.

Procedure step-by-step:

- Dismantle the shoring that supported the slab.
- Bring excavated soil and backfill in layers in the trenches, compacting up to elevation –0.565 m, up to the locations of foundation plots containing service sleeves (water/sewer) (Figure 42q).
- Excavate trenches from the foundation faces where sleeves are set to the exit points through the slab at –0.05 m, within the interior of the house.
- Remove the sleeve caps and install the sewer pipes piece by piece with direction-change fittings to the exit point above the ground-floor slab; bring the PPR water pipe likewise to its indicated location.
- Seal the ends of both utility lines, then backfill and compact the trenches.

5.4.8 Final Strengthening of TFM Walls at the Bottom Part

This part can also be replaced with a solution chosen from Chapter 4 (based on natural materials), depending on the location of the building (seismicity), the engineer's recommendation and also owner preference. For example, the TRAROM solution (Chapter 4.4.1 above) works well in this situation, as long as the timber frame is healthy or replaced as described in the steps above, so the diagonal planks can be connected to it.

Quick glossary (what's what):

- L-shaped meshes at corners — Bent mesh pieces that wrap around outside/inside corners so reinforcement remains continuous.
- Connecting ties (clips) — Short steel anchors set into drilled holes that secure the mesh to the masonry substrate.
- Vertical starter bars (from plinths) — Short vertical rebars protruding from the concrete plinth/foundation, used to tie the new jacket into the base.

Welded meshes shall be installed along the wall bases, around the full perimeter and on both faces (Figure 43).

Procedure step-by-step:

- Set out the positions of mesh panels at the wall's bottom part around the full perimeter, ensuring 3-mesh overlaps in both directions, in plane and at corners (L-shaped meshes).
- Mark the locations for connecting ties (clips) anchoring the meshes at the wall bottom part (according to the project).
- Drill channels/holes along the panel lengths without vibrations.
- Insert the ties and grout them after properly wetting the surrounding masonry.
- Cover tie ends with masking tape.
- Wet both faces along the wall bases and apply a uniform spatterdash bonding coat (with additives).
- Remove the masking tape.
- Bring welded mesh panels (sized per design) onto the primed masonry at the wall bottom parts; support them on the tie ends and over the vertical starter bars protruding from the plinths.
- Fix the meshes by bending the tie ends vertically over the mesh nodes. Install local PVC spacers to allow mortar to pass between wall and mesh.

Note: If this wall meets a RC tie-column (e.g., at the edge of the one-brick envelope wall at the intersection between axes A and 4, or between axes A–B and axes 4–5), then after fixing meshes on both faces, install pairs of horizontal starter bars (projecting bars) to connect the jacketed TFM wall edge to the RC tie-columns (per project).

5.4.9 Plastering the Bottom Part of TFM Walls

The plaster can be any type of material (clay or cement based, depending on the engineer and owner's choice). Nowadays material manufacturers provide us with many types of materials, nature based, that can meet current requirements.

Quick glossary (what's what):

- Straightedge — A long, rigid ruler/section (aluminum/wood) used to check plane and to level/strike off fresh plaster.

Procedure step-by-step:

- Mark plaster thickness with mortar beacons at approximately 1.50 m horizontal and 0.50 m vertical spacing down to the wall base, checking coplanarity with a straightedge (relative to the already plastered surface). Then fill vertically between beacons (after proper wetting), forming parallel vertical guides ensuring constant plaster thickness.
- Wet the surfaces to be plastered, then apply mortar uniformly on both faces. The mortar shall gradually penetrate behind and embed the fixed meshes, from

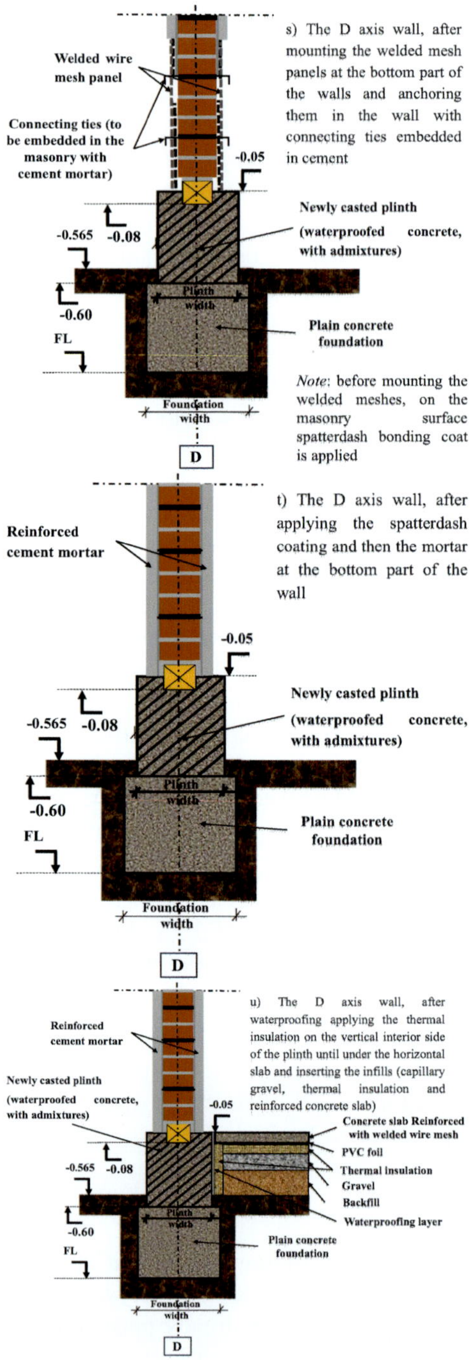

FIGURE 43 Strengthening the TFM walls at the bottom part and connecting it with the new plinth's reinforcement.

the previous plaster stop level down to the wall base. Level with a straightedge drawn downward along the vertical guides.

5.4.10 Demolition of Perimeter Walls (One-Brick Thick) Surrounding the Entrance Hall

Quick glossary (what's what):

- Rough-sawn timber strips (≈ 7 × 5 cm) — Unplanned/unprocessed wood laths about 70×50 mm, formerly nailed under the slab to support infill/formwork.
- Infill between slab beams — Non-structural material (e.g., planks, light concrete, hollow units) placed between beams; removed to access the soffit.
- Joinery — Doors, windows, and their frames set in the walls.
- Pipe columns — Steel tubular columns carrying the entrance slab; removed after the slab is down.

Procedure step-by-step:

- Remove the remaining rough-sawn timber strips (≈ 7 × 5 cm) nailed to the soffit of the slab in the bay between axes A/B/4/5, after removing the infill between slab beams.
- Set out alignment beam positions at ground level –0.565 m for the shoring props supporting the floor in the A/B/4/5 bay.

Note: When positioning alignment beams, consider the locations and sizes of the envelope wall foundations to be built.

- Place alignment beams on the markings.
- Erect the shoring supporting the slab above the entrance hall (properly braced); set props on alignment beams; slab beams are bearing on props via perpendicular crosspieces.
- Disconnect the connections between slab beams and the timber top plate at the top of the envelope wall on axis A between axes 4 and 5, and the upper portion of the TFM wall on axis B between axes 4 and 5.

Note: If required to stabilize the wall on axis B during demolition of the adjoining envelope walls, connect it to the erected shoring.

- Provisionally stabilize the envelope walls flanking the entrance hall (by exterior propping).
- Slightly jack the slab beam ends (above the hall) until contact is lost with the timber top plate sitting on the wall on axis A between axes 4 and 5.
- Remove joinery in the one-brick envelope walls along axis A between 4–5 and along 5 between A–B.
- Perform controlled top-down demolition of these one-brick envelope walls, carefully minimizing vibrations and shocks.

Note: During this operation, also demolish the RC slab above the main entrance, bearing with its outer edge on two pipe columns.

- Demolish the entrance stairs and remove the pipe columns supporting the slab.
- Demolish the unreinforced-concrete plinths belonging to the above envelope walls.
- Remove debris and level the ground surface in the area (–0.565 m).

5.4.11 Execution of an Unreinforced-concrete Foundation and a RC Plinth for the New Structural Walls (Confined Masonry) along Axis A between Axes 4–5, and along Axis 5 between Axes A–B

Quick glossary (what's what):

- Batter boards & nails → temporary timber frames that hold axis wires (string lines).
- Epoxy-anchored dowels → rebars bonded in drilled holes to lap the new reinforcement.

Procedure step-by-step:

A) Marking of axes and foundations
   - Mark axes A and 5 as extensions of the existing TFM walls.
     How/Where: Fix axis ends to nails in timber posts and to timber batter boards set at practical offsets.
   - Stretch axis wires between the ends and check orthogonality.
     Why: Ensures right angles at the grid.
   - Mark the foundation lines for the envelope walls on the ground at (–0.565 m), referenced to axes A (4–5) and 5 (A–B).
     What: Paint/chalk the trench footprints.
   - Verify the marking, then remove the axis wires (so excavation can start).

B) Excavation and screed guides
   - Excavate trenches for the envelope-wall footings to design depth; dispose of spoil.
     Note: As per 5.4.7.3.
   - Check trench geometry (widths, depth, straightness).
   - Fix horizontal screed planks around trench perimeters at (–0.565 m) and pin them to the ground.
   - Level the planks (use wedges if needed) and verify their elevation vs benchmark (±0.00).
   - Mark on plank tops the positions of the starter-bar cages (for later tie-column cages), per drawings.
   - Re-stretch axis wires A and 5 between benchmarks and re-check orthogonality (control check).

C) Place starter cages and cast the continuous footing
   - Lower the starter-bar cages into the trenches; hold each vertical using pairs of parallel battens crossing the cage (let one stirrup rest on the battens) and support on the marked planks.
     Requirement: Longitudinal bars must penetrate below plank level to the design embedment.
   - Center cages to the axes and fix battens to the planks.
   - Then remove axis wires.
   - Place and compact concrete in the foundations, without disturbing cage positions.
   - Level the concrete surface to (–0.60 m) using a profiled straightedge resting on the planks.
     Note: Use waterproofing admixture in this continuous-footing concrete.
   - After initial set, remove the cage-support battens and planks; cure the footing (keep moist by watering or wet mats).
   - ~24 h after casting, remove hardened laitance from vertical starters and clean cement film/loose fines from the horizontal surface where the plinth will be laid out.

D) Prepare old plinth ends and re-establish axes
   - Pick-hammer (bush-hammer) the end faces of existing TFM envelope-wall plinths (from the demo of the one-brick portions) along A (4–5) and 5 (A–B) to obtain flat faces.
   - Clean the continuous-foundation surface.
   - Re-stretch axis wires A and 5 and check orthogonality.
   - Transfer axes onto the foundation with a plumb; mark ends.
   - Snap chalk lines of the axes on the foundation surface.
   - Mark the new RC plinth footprint on the foundation relative to the axes.
   - Remove the axis wires (formwork/rebar work begins).

E) Drill, dowel, and place the new plinth cages
   - From the strengthening design, determine positions of drilled holes in the old plain-concrete plinth ends for rebar dowels/starters that will lap the new plinth longitudinal bars.
   - Drill to design depth/diameter and clean the holes.
   - Install reinforcement cages for the new plinth (along axis A and along axis 5).
     Detail: The vertical starter bundles for tie-columns must pass between the plinth's longitudinal bars.
     Add spacers (sides and bottom) for cover.
   - Verify cage positions and adjust as necessary.
   - Fix rebar dowels with epoxy mortar in the drilled holes at the old plinth ends, lapping the longitudinal bars of the new plinth cages.

F) Form, pour, and cure the new RC plinth
   - Install side formwork for the new plinth; verify the top edge elevation (–0.05 m) relative to (±0.00).
   - Pre-wet timber formwork ~2 h before casting; remove any pooled water at the base before the pour.

- Pour and compact concrete in the new plinth along A (4–5) and 5 (A–B); level with a straightedge along the formwork top.
- After ≥24 h, remove hardened laitance from the protruding starter bars.
- When concrete reaches the required maturity, strip side forms and treat/cure exposed concrete surfaces.

Element map (where each thing sits):

- Trenches & continuous footing: along axes A and 5; top at ≈ −0.60 m after leveling.
- Starter-bar cages (footing): aligned to axes; held by battens on screed planks during the pour.
- Old plinth ends (TFM walls): bush-hammered to flat; receive epoxy dowels that lap into the new plinth cages.
- New RC plinth: cast above the footing within side forms; top at −0.05 m; carries tie-columns and masonry.
- Axes & control: set by batter boards, nails, and wires; repeatedly checked for orthogonality and elevation.

Tips:

- Recheck orthogonality and levels after every setup change.
- Keep cover blocks in place; don't displace cages during concreting.
- Strictly clean drilled holes before epoxy dowels.
- Maintain curing to specification; remove laitance where noted to ensure good bonds.

5.4.12 Execution of Structural Walls with Confined Masonry along Axis A (4–5) and Axis 5 (A–B)

Confined masonry is not always RC confined masonry. It can also be, as the existing system of this house (most of it), TFM, which means the confining of the masonry is done by the timber frame. As also mentioned before, further we give the technological steps on how to make RC elements to confine the masonry, being the most common solution selected, but in fact, if chosen, the timber frame is easier to install and less time consuming.

Quick glossary (what's what):

- Tie-beam (RC beam/ring beam) → RC beam at the top of the wall, tying columns horizontally.
- Lintel → small RC/timber beam over an opening (window/door).
- Connectors → horizontal steel bars tying existing walls (to be jacketed with reinforced plaster) to RC columns.
- Inspection/pouring windows → small openings in column formwork every ~0.90 m to help pouring and compaction.

- Processed/dressed timber members → planed, dried, treated wood placed on tie-beams as bearings for slab beams.

Procedure step-by-step:

A) Markings and column cages
   - Mark door opening and tie-column positions on the top surface of the new plinth.
     Why: Accurate layout before any steel or masonry goes in.
   - Install tie-column cages.
     What: Longitudinal bars continue from the plinth starter bars; place the lower bundles of stirrups to the marked levels and tie, with denser stirrups over the splice height.
     Note: If needed, temporarily brace the vertical cages to nearby shoring (supporting the slab) in bay A/B/4/5.
   - Install horizontal connectors between the existing walls to be jacketed and the tie-column cages at intersection of axes A and 4 and at intersection of axes B and 5 (see Figure 38b).
     Why: Ensures composite action between jacketed masonry and RC frame.

B) Masonry infills (up to sill level, then to heads)
   - Build one-brick masonry infills (solid bricks 240×115×63 mm) between tie-columns, on the marked lines, up to the windowsill level (where applicable).
     How: Place specified horizontal reinforcement in the bed joints and anchor it into the tie-column cages.
   - Check the masonry for plumb and level.

Important execution notes:

- The first bed joint on the plinth uses waterproofed mortar (rigid waterproofing). Alternatively, apply liquid waterproofing on the plinth before laying the first joint.
- At column interfaces, tooth the vertical edges of the masonry panels for good bond with the jacket/columns.
- Wet bricks properly; lay them after surface drying.
- Respect bonding rules and fully fill joints.
- Materials (brick/mortar/concrete/rebar) per the strengthening project.
- Mark window positions at mudsill level.
- Continue the one-brick masonry, alternating with reinforcement in the specified bed joints, from opening jambs toward tie-columns and the door opening, up to the head of each opening.
- Mark lintel seat positions and lengths above openings.
- Build courses from outside the lintel seats up to the underside of the tie-beam level (leave the lintel gaps open).
- Re-check masonry for plumb and level.

C) Tie-columns: cover, formwork, and concrete
- Install spacers on column cages to secure concrete cover.
- Install column formwork, leaving inspection/pouring windows every ~0.90 m vertically on one face.
- Pre-wet the toothed masonry and timber formwork ≥2 h before concreting; at pour time surfaces must be damp, not wet. Drain excess water through the windows.
- Close the lowest windows and pour concrete in ~30 cm lifts; compact by light hammering of the formwork with a ~2 kg wooden mallet and by rodding.
  Aim: Avoid air pockets under any cantilevered parts.
  Sequence: Start through the intermediate windows, then pour from the top.
  Stop level: ~3 cm below the top face of the last laid brick course (for a good tie-beam bond).
- After ≥24 h, strip the vertical faces of the column formwork; cure the exposed concrete and clean hardened splashes and cement film from reinforcement just above the pour stop.
  Curing note: Keep RC surfaces moist for ≥7 days (plain Portland) or ≥14 days (blended cements).

D) Lintels and tie-beams (reinforcement, formwork, concreting)
- Install soffit/bottom formwork for lintels over windows/doors.
- Install reinforcement cages for tie-beams and lintels
  - The tie-beam cage passes through the lintel cage.
  - At axis 5, provide anchorage length ≥ 60ϕ into the perpendicular tie-beam.
  - Support cages on masonry/formwork with spacers for cover.
  - Also place starter bars for the processed timber members that will sit on the tie-beams (as bearings for slab beams).
- Install side spacers on tie-beam and lintel cages.
- Install side formwork for tie-beams and lintels; ensure starter bars are plumb and project above the formwork to the required height.
- Pre-wet timber formwork and the masonry contact surface ≥2 h before the pour.
- Pour and compact the tie-beam concrete.
- After ~48 h, strip the vertical faces of lintel and tie-beam formwork and cure exposed concrete.
  - Keep the soffits/bottoms in place until the slab (currently shored) can safely bear.

E) Waterproofing at tie-beams & timber bearing
- Prime the horizontal tie-beam top with deep-penetration bituminous primer, then adhere bituminous waterproofing strips.
- Place dressed, dried, antiseptic-treated timber members along the tie-beams (as bearings for slab beams) and fix/confine them with the starter bars projecting from the tie-beams.

Element map (where things sit)

- Plinth (base RC band): at floor level; carries tie-column starters.
- Tie-columns: rise from the plinth at grid points and jambs; later jacketed walls tie into them via connectors.
- Masonry infills: one-brick-thick panels between columns, toothed at edges; reinforced in specified bed joints.
- Openings: lintels span over windows/doors; built to the specified seat and length.
- Tie-beam (ring beam): at the top of the wall, continuous through lintel zones; receives starter bars for timber.
- Waterproofing: on the horizontal tie-beam top (primer + bitumen strips) beneath the timber bearing.
- Timber members: laid on tie-beams, fixed by starter bars; later carry the slab beams.

Tips:

- Check layout before fixing steel.
- Maintain cover with spacers; keep formwork windows clean.
- Respect pour heights and curing times.
- Keep starter bars straight and to height for proper timber fixing.
- Verify plumb/level of masonry after each lift and before concreting adjacent RC.

5.4.13 Investigation and Replacement of Deteriorated Timber Beams in the Slab above Ground Floor

Quick glossary (what's what):

- Walking boards → temporary boards for safe access at the soffit
- Tie beam/bond beam → horizontal beam tying the wall head; top plate (timber) sits on the wall
- Strap plates/fishplates → steel (or timber) side plates bolted to the beam to join a new segment
- Anchors/metal angles → connectors fixing beams to wall top plate or to dressed timber on tie beams

Procedure (step-by-step):

- Bring walking boards to the soffit in the jacketed-wall zone; inspect beam soffits.
  Where: Under the slab, near jacketed walls.
  What/Why: Add safe access; start a visual check for defects.

Action: Remove some rough-sawn strips (~7×5 cm) from the soffit to open inspection space.

- Entrance hall: plank under the current shoring; inspect beams; remove more strips if needed.
  Where: Entrance-hall zone under the slab.
  Why: Improve bearing of shoring and expose beam undersides for assessment.
- From above, inspect the slab carefully; remove old deck boards if needed.
  Why: To see beam tops and detect hidden decay around bearings/notches.
- Mark beams with deteriorated zones.
  What: Chalk/paint lines indicating defect extents.
- Entrance hall: erect initial shoring for slab beams, bearing on the wall along axis A (4–5).
  How/Where: Use processed timber members fixed on tie beams; then reconfigure shoring on both sides of the damaged beam for localized support.
  Why: Create a safe load path before any cutting.
- In the area bounded by jacketed walls: set out at (−0.565 m) the positions of alignment beams for shoring beneath slab portions between damaged-beam zones.
  Action: Install braced props up to the beam soffits, each prop bearing via perpendicular crosspieces.
  Why: Continuous, well-distributed temporary support.
- Remove all rough-sawn strips (~7×5 cm) under beams along their length, and remove the top deck boards; disconnect old ties from wall top plates.
  Why: Full access to beam sides and bearings.
- Depending on defect position, prop the healthy beam portions.
  How: Use carpentry methods, adjacent shoring, or props sitting on alignment beams.
  Why: Maintain stability while the damaged part is removed.
- Mark cut lines and measure the clear replacement length.
  Where: Beyond the visibly deteriorated wood, into sound timber.
- Cut out the damaged portions.
  Safety: Cutting only after shoring is verified.
- Bring and install new segments (sound, dried, antiseptic + fire-retardant treated timber).
  How: Lateral steel strap plates and bolts per design (or timber fishplates if specified).
  Where: Plates span across the joint, on both sides.
- Re-support the beams.
  What: Gradually transfer load back from props to repaired beams; adjust shoring.
- Check all beam supports; remove any remaining deteriorated deck boards.
  Note: Slab stability is ensured by the installed shoring and any soffit strips retained for bracing.
- Replace damaged anchors.
  How/Where: Install metal angles screwed to the beam and to the wall top sill or to dressed timber on bond beams.

Why: Restore proper wall-to-beam connections.
- Treat beam sides and tops with antiseptic and fire-retardant spray.
  Where: All exposed timber, including new splices and adjacent existing wood.

Element map (where each piece sits):

- Beams: run under the deck boards; you can see their soffit from below and top from above.
- Rough-sawn strips (~7×5): nailed to the soffit; removed to create workspace.
- Shoring props: stand on the floor or on alignment beams (–0.565 m); top ends push to the beam soffits via crosspieces; braced laterally.
- Jacketed walls: bound the work area; provide safe supports for shoring/tie-ins.
- Tie/bond beams + top plates: at wall heads; anchors/angles connect beams here.
- Strap/fish plates: fixed on beam sides across the new splice joint with bolts.
- New timber segment: replaces the cut-out zone; same section/species as design project.

Tips:

- Verify shoring capacity and bracing before any cuts.
- Extend cut lines into sound wood; ensure dryness of new timber.
- Tighten bolts to specified torque; maintain edge distances.
- Record as-built splice lengths, plate sizes, and treatment batch.

5.4.14 Installation of a New Timber Slab, Two Overlapping Plank Layers at 45° to the Slab Joists Axes

Quick glossary:

- Joist— horizontal timber beam of the floor
- Thick plank/board (decking board) — laid in two layers at ~45° to the joists (between 2.5 and 4 cm thickness)
- Blocking/nogging between joists — short pieces of timber planks set between joists, often at wall lines
- Steel angle seat (L-bracket) — fixed to the sides of joists to support plank ends
- Jack — mechanical device used to lift heavy objects by applying a small force at one point and generating a much larger lifting force, in this case used to gently take load and hold a member temporarily
- Rolled U-channel — temporary beam under a post, sitting on jacks
- Steel cleats/plates on the post — to temporary lift each roof post so you can shorten it to the new floor level.
- Roof post — vertical post that carries the roof load down to the floor
- Post base plate/sole plate — what the roof post sits on, on top of the floor

Procedure step-by-step:

Stage 1 – Preparing the structure

- Unload the roof frame.
  What: Remove part of the roof covering.
  Why/Where: Reduces weight on the roof posts that currently bear on the floor you're replacing.
- Install shoring (Figure 44a).
  What: Put props under the joists to support the whole floor during the work.
  Where: From the space below, directly under joists; connect into a small scaffold if needed.
- Add blocking between joists (Figure 44a).
  What: Fit blocking pieces (treated against fire/decay) between joists, aligned with the walls. Screw them in.
  Why/Where: They become the first supports for the ends of the new planks near the walls.
- Fix steel angle seats.
  What: Screw L-brackets (steel angle seats) to the sides of the joists (Figure 44a).
  Where: Set their top flange one plank thickness below the joist top — this lines up the first decking layer.

Stage 2 – Building the support for the decking

- Short planks between joists.
  What: Place short, horizontal planks on those angle seats, between joists.
  Why/Where: They double the blocking and create a continuous support frame.
- Add intermediate supports.
  What: Similar short planks at mid-span and at the quarter points between joists.
  Why/Where: Gives uniform support for the decking across the bay (also around the attic hatch).

Stage 3 – Laying the first decking layer (at 45°)

- Cut at 45°.
  What: Prepare long planks with 45° end cuts so they meet the edges neatly.
- Lay the first layer (Figure 44b).
  What: Start from a corner and lay planks at ~45° to the joists.
  Where: Planks bear on joist tops and on the intermediate supports you just made.
- Screw everything down.
  What: Fix each plank with wood screws into the joists and supports.

Stage 4 – Working where roof posts sit on the floor

- Fit cleats to the roof post (Figure 44c).

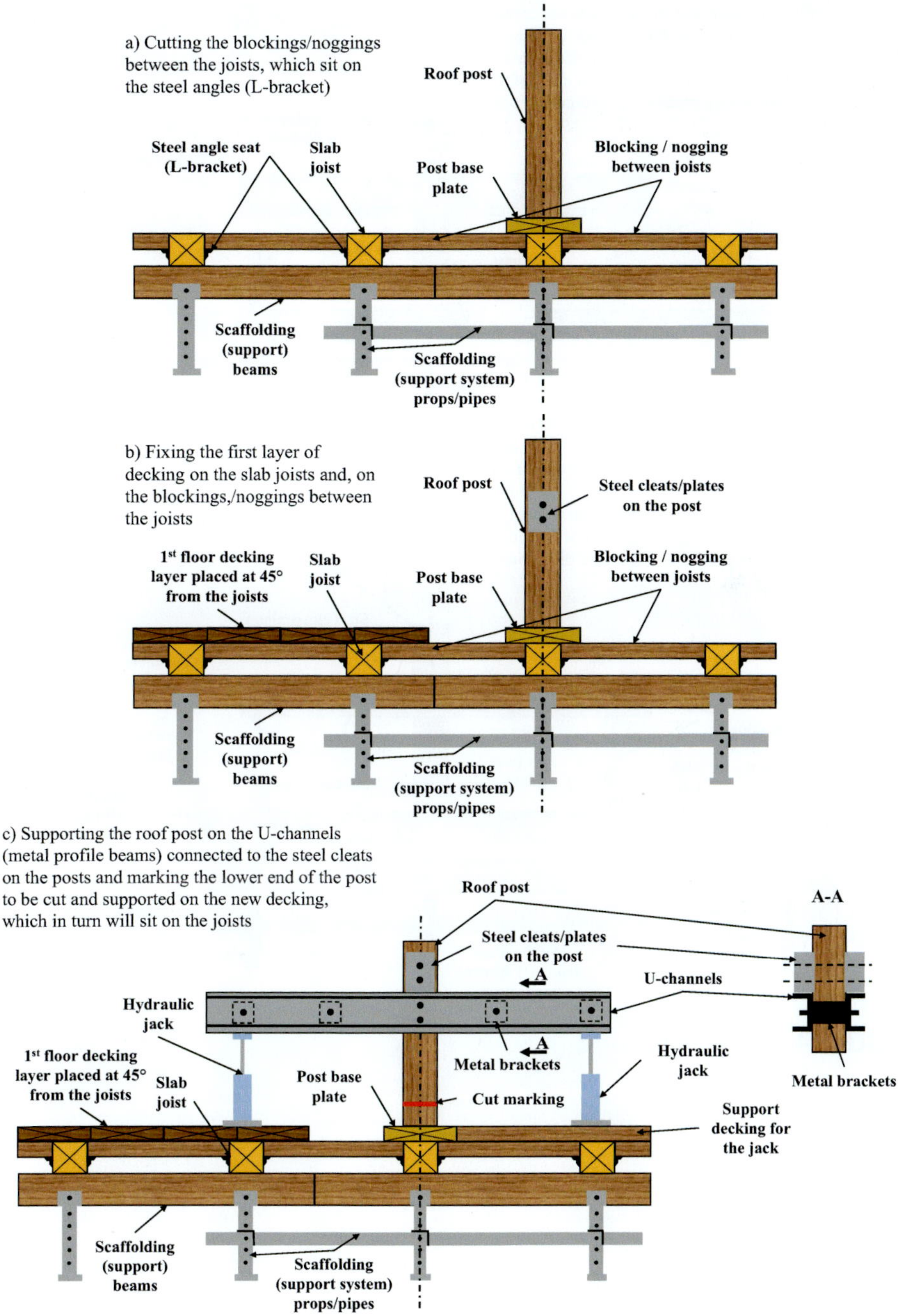

FIGURE 44 Installation of a new timber slab, two overlapping plank layers at 45° to the slab joists axes.

d) Cutting the lower end of the post to the correct level to be supported on the new decking to be finished (2 layers of decking, laid at 45° from the joists' axis and transfer the load from the roof to the U-channels supported by the hydraulic jacks

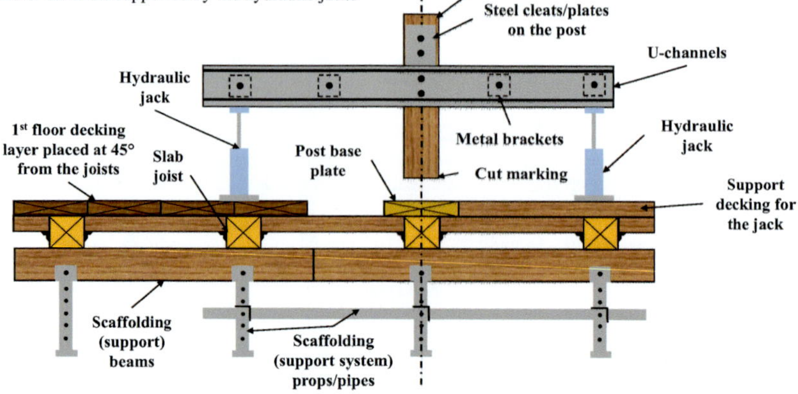

e) Refixing the post base plate at its newly cut lower end and finishing laying the new decking (layer 1) on the slab joists and on the blockings between the joists

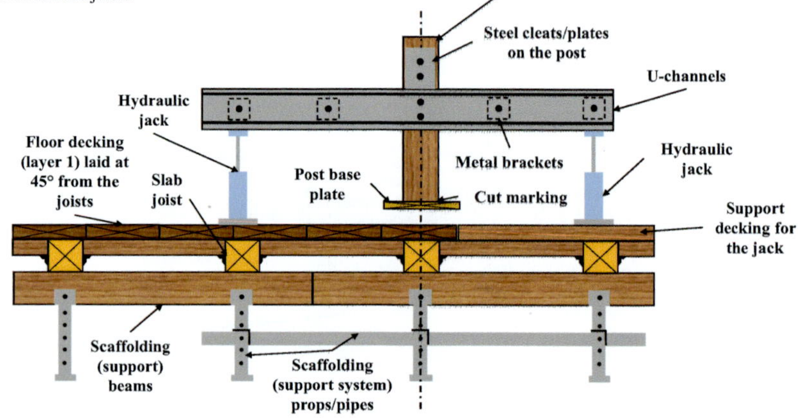

FIGURE 44 (Continued)

f) Adding a support decking piece between the post's base plate and the decking layer 1, lowering the hydraulic jacks and removing the U-channels
Note: The load from the roof framework is now carried by the posts and transferred to the slab by means of this added piece.

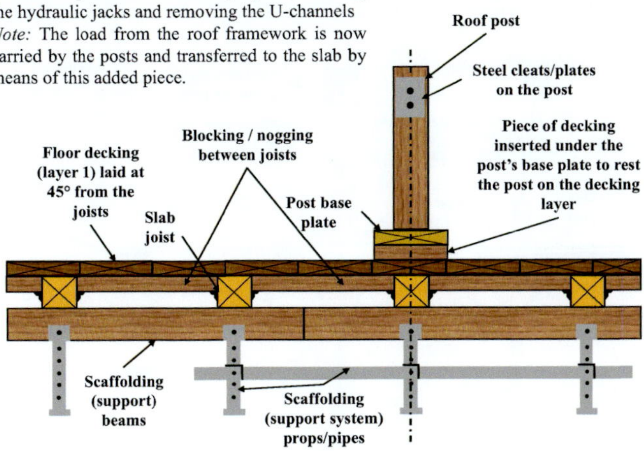

g) Starting to fix the second layer of decking, at 45° angle from joists' axis, directly on the first layer
Note: The angle between the axis of the 2nd layer of decking to the axis of the 1st layer of decking is 90°

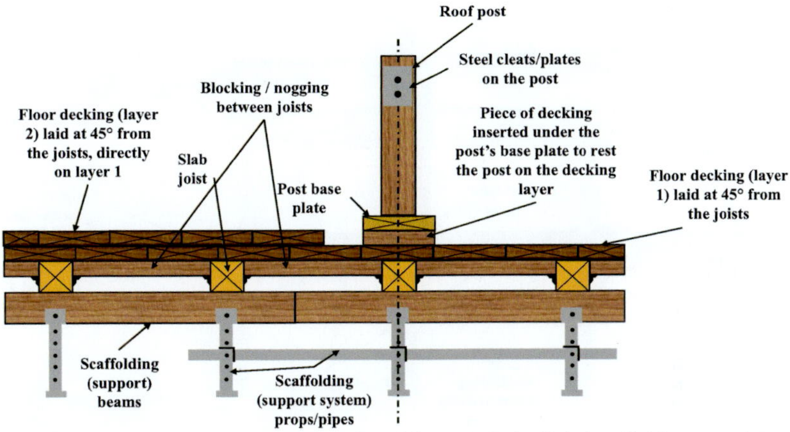

h) Re-installing the U-channels to support the roof post, these sitting on the hydraulic jacks, which in turn are sitting on the 2 layers of newly made floor decking
Note: By easy lifting the jacks, the load carried by the post is taken by the U-channels, allowing the removal of the piece of decking under the post's base plate and on the 1st layer of new decking

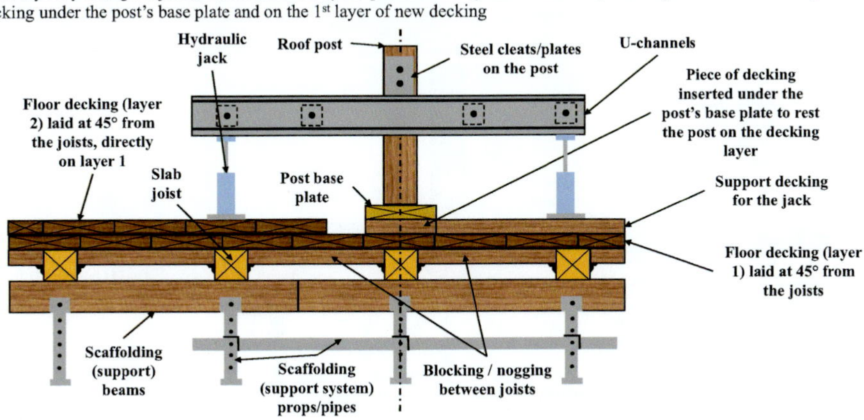

FIGURE 44 (Continued)

i) Removing the piece of decking inserted under the post's base plate to rest the post on the decking layer, ensuring the possibility of mounting further the decking of the 2nd layer

Note: The angle between the axis of the 2nd layer of decking to the axis of the 1st layer of decking is 90°

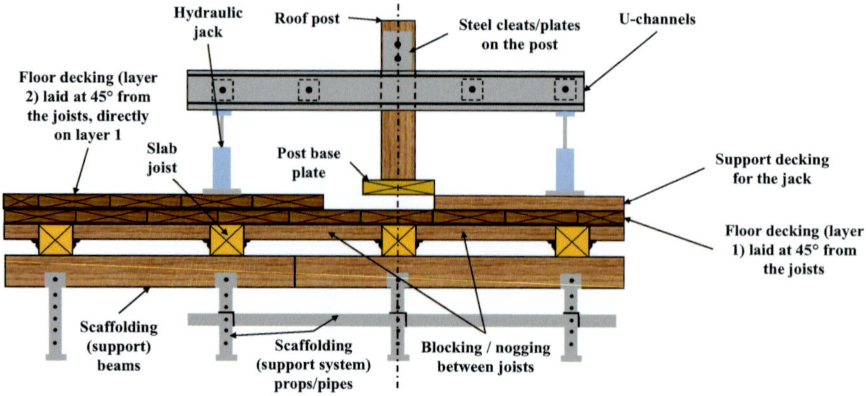

j) Mounting further the decking of the 2nd layer, under the roof post base plate

Note: The angle between the axis of the 2nd layer of decking to the axis of the 1st layer of decking is 90°

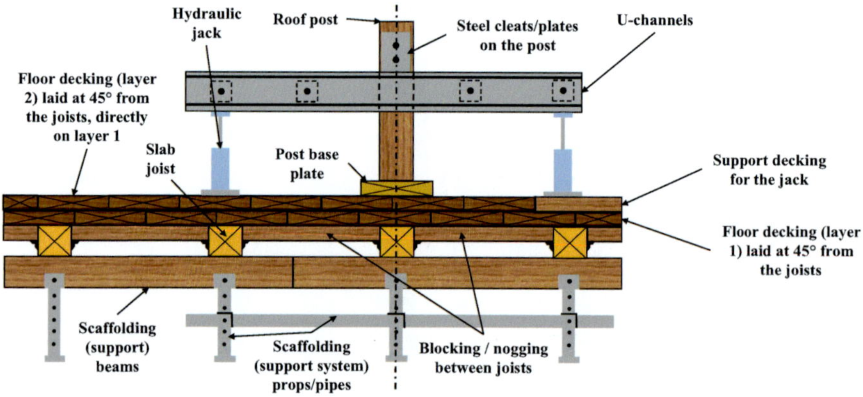

k) Removing the U-channels and the steel cleats/plates that supported the roof post, after finishing mounting of layer 2 of decking

Note: After finishing the mounting of the 2nd layer of decking, the bottom part of the roof posts are fixed on the base plate, and in turn, the base plate is fixed to the decking.

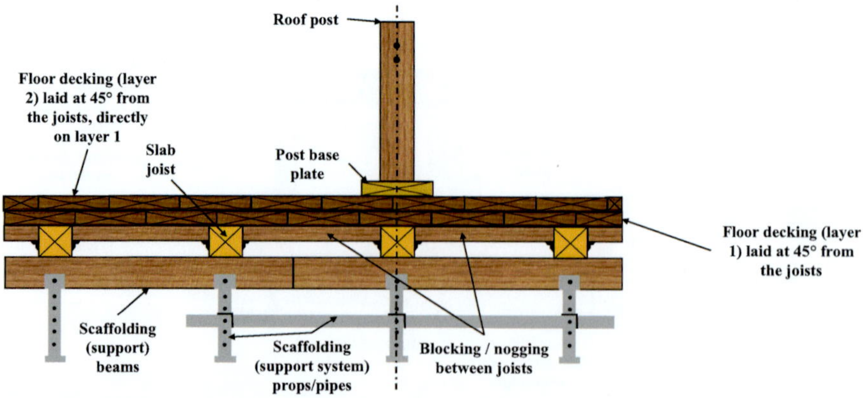

FIGURE 44 (Continued)

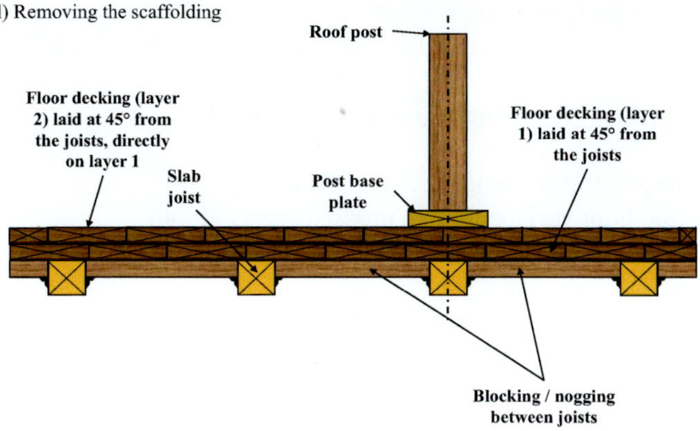

FIGURE 44 (Continued)

What: Attach steel plates/cleats to the roof post so you can pick it up temporarily.

- Add a U-channel and jacks (Figure 44c).
  What: Slide a rolled U-channel under those cleats; support the channel with two jacks on solid points (props/shoring).
  Why/Where: This bypasses the floor, carrying the roof post load down to the temporary shoring.
- Gently lift with the jacks (Figure 44d).
  What: Raise a little to take the load off the post's base.
- Shorten the roof post.
  What: Cut the bottom of the post so its new length is shorter by the total thickness of the two decking layers.
- Re-fix the post base (Figure 44e).
  What: Put the base plate back under the post at the new height (on top of the new decking level).
- Continue decking under the post (Figure 44e).
  What: Keep laying planks past the post location until you're clear of it.
- Lower the jacks and remove the U-channels (Figure 44f).
  What: Let the load return to the post resting on its new base; take away the U-channel and jacks.
- Repeat this sequence for every roof post you encounter until the whole area is decked.

Stage 5 – The second decking layer

- Treat the first layer.
  What: Preservative + fire-retardant spray on the first layer.
- Lay the second layer (Figure 44).
  What: Install the second set of planks also at ~45° but in the opposite direction, making an X with the first layer.
- Final fixing.
  What: Screw the layers securely; apply a final treatment if specified.

- Fix the roof posts to the new floor.
  What: Bolt/screw each post base down to the finished decking.

Simplified "who does what, where" recap:

- Joists: the structural frame; decking sits on top.
- Planks: the floor surface, in two crossed layers at ~45°.
- Blocking & short between-joist planks: edge and intermediate support between joists.
- Angle seats (L-brackets): metal shelves on joist sides that hold plank ends.
- Shoring, props & jacks: temporary support carrying loads during the works.
- U-channels & cleats: the temporary lifting system for each roof post so you can shorten it to the new floor level.

In the end it results a new, solid floor deck of two 45° layers, well supported on the joists. Roof posts correctly re-seated at the new height and fixed down. The whole assembly is stiffer, safer, and more durable.

5.4.15 Execution of Slab Infill and Layers for the Interior Slab System

Quick glossary (what is what):

- Interior shoring — Temporary supports inside the building that hold elements in place during work; removed once the structure is stable.
- Deep-penetration bituminous primer — A thin, low-viscosity asphalt-based liquid that soaks into concrete surfaces to improve adhesion and water resistance for subsequent waterproofing.
- Brush-on waterproofing — Liquid membrane applied with brush/roller that cures to form a continuous water barrier.
- XPS boards — Extruded polystyrene rigid foam boards; closed-cell insulation with low water absorption, used for thermal protection.

Procedure step-by-step:

- Dismantle interior shoring.
- Apply deep-penetration bituminous primer to the interior vertical faces of the RC plinths under the wall system, up to the soffit of the RC slab with top at (–0.05 m).
- Apply brush-on waterproofing over the primer.
- Protect the interior plinth faces with XPS boards up to under the slab (top at –0.05 m).
- Bring excavated soil and spread between plinths; compact in layers; raise utilities as needed (sewer), keeping pipe sections in position.
- Place a uniform capillary gravel layer over the compacted fill.
- Install thermal insulation according to the strengthening project's thickness (i.e.: XPS or a natural one), cutting around protruding pipes.
- Lay PVC foil over the insulation.

- Reinforce the slab at (–0.05 m) with welded mesh panels overlapped 3 meshes in both directions (mesh cut around pipes).
- Pour and compact the slab concrete; at the proper time, take measures to prevent moisture loss.

5.4.16 Construction of Exterior Stairs, Perimeter Walkway, and Plastering of the Exterior Plinth

Quick glossary (what is what):

- Subgrade – The compacted soil layer that supports foundations or slabs.
- Elevation – The height or vertical level of a specific part of a structure.
- Slope (fall) – The intentional incline of a surface to direct water away.
- Riser – The vertical face of a step.
- Tread – The horizontal surface of a step that you step on.
- Landing – A flat platform area between flights of stairs.
- Footprint/set out – The exact layout or outline of a structure on the ground before construction
- Kerb (curb) – The raised edge or border between a sidewalk and roadway or ground.
- Ballast – Coarse granular material (e.g., gravel) used for stability under kerbs or slabs.
- Foil-wrapped plywood edge – Temporary smooth formwork covered with foil to prevent bonding and ease removal.
- Bay ("plot") – A divided section of a large concrete area to control cracking and construction sequence.
- Joint filling (bitumen) – Sealing the gap between concrete bays with a flexible, waterproof material.

Procedure step-by-step:

A) External stairs (foundations → steps)
   - Mark the stair foundations position. Its width should allow the manual digging (approx. 45 cm) (Figure 45b)
     What: Mark the exact footprint/axes.
     Where: On the ground next to the building.
     Note: The width adopted for the foundation trench must ensure the proper bearing of the last two steps of the stair on the surface of the first step, which is cast together with the foundations.
   - Excavate trenches for the stair foundations to the design level (Figure 45c).
     Where: Immediately alongside the building foundation.
   - Clean the exposed face of the building foundation (in the stair area).
     Why: For good bonding of dowels and concrete.
   - Making the formwork of the first part of step 1. The vertical edge of the first step is done with approximately 10 cm lower than the final level of step 1 (–0.285) (Figure 45d).

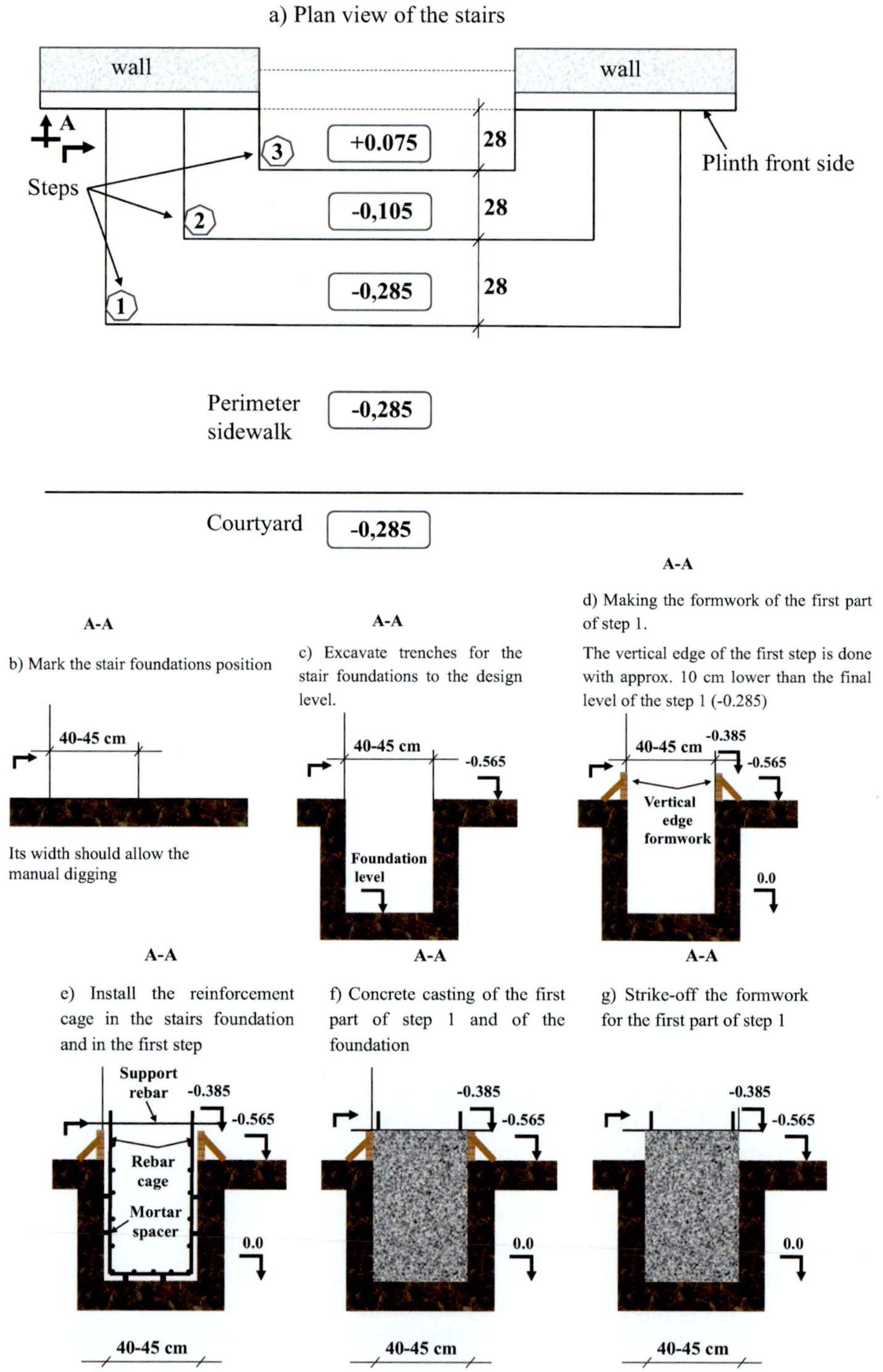

FIGURE 45 Steps in making exterior stairs.

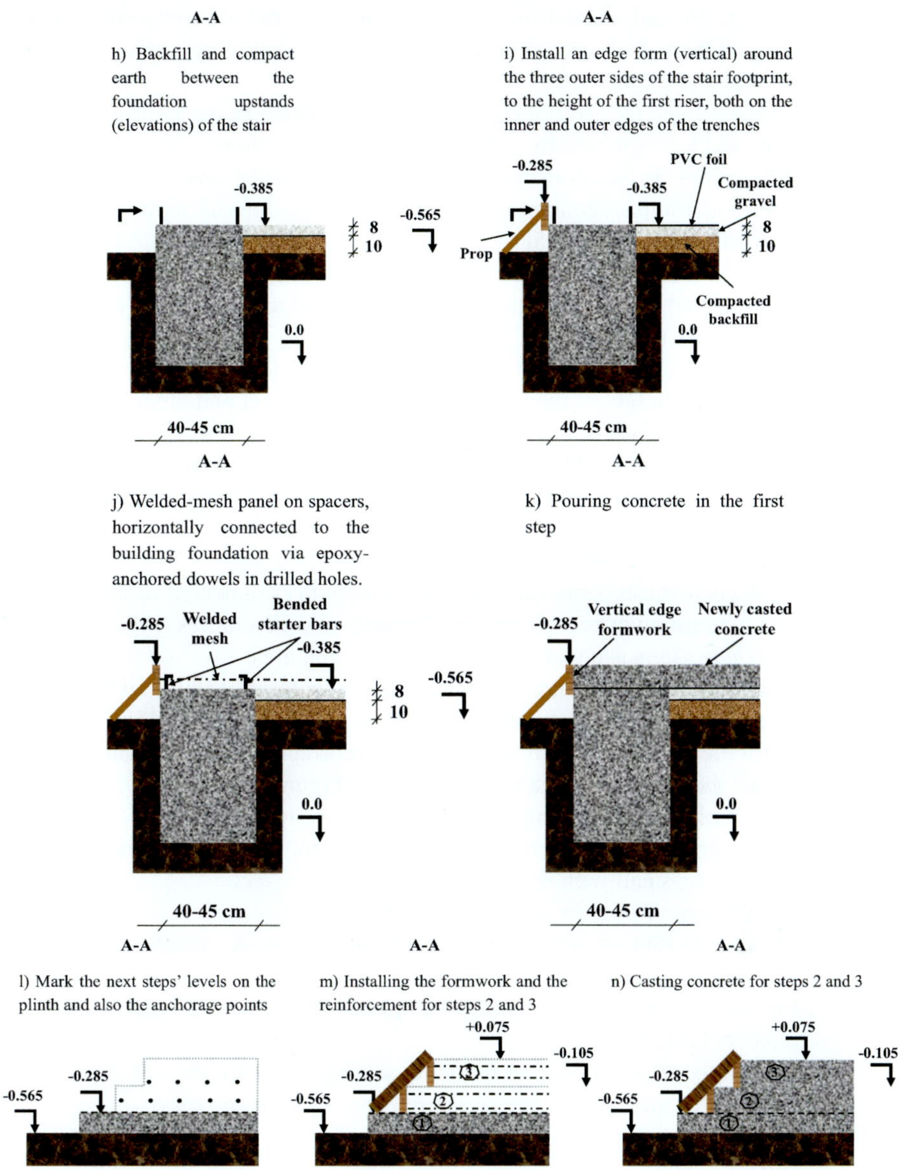

FIGURE 45 (Continued)

- Install rebar dowels with epoxy grout in drilled holes on the outside face of the building foundation, at the spots where the stair foundation reinforcement will connect.
  Why/Where: These tie the new stair foundation to the existing foundation.

- Place U-shaped welded-mesh cages in the three outer foundation trenches; support them with mortar spacers (blocks tied with wire) under and along the sides (Figure 45e).
- Connect the cages to the epoxy-anchored dowels in the building foundation; add L-shaped mesh corner pieces at intersections.
  Where: This is the reinforcement of the stair foundations.
- Pour and compact concrete into the stair foundations up to the top edge of the formwork (Figure 45f).
- Strip the form of the first step at the right time (Figure 45g).
- Backfill and compact earth between the foundation upstands (elevations) of the stairs (Figure 45h).
- Install an edge form (vertical) around the three outer sides of the stair footprint, to the height of the first riser, both on the inner and outer edges of the trenches.
  Why: Forms the visible outer curb of the first step (Figure 45i).
- Lay a capillary break (Figure 45i) and slab reinforcement (Figure 45j):
  - Capillary gravel layer (Figure 45i),
  - PVC sheet (Figure 45i),
  - Welded-mesh panel on spacers, horizontally connected to the building foundation via epoxy-anchored dowels in drilled holes (Figure 45j).
  - *Where:* This is under the first tread/landing area.
- Pour and compact concrete for the first stair tread/landing and strike off level with the tops of the stair foundation upstands (Figure 45k).
- Mark the next steps' formwork up to the correct elevation (Figure 45l).
- Install the step formwork up to elevation (Figure 45m).
  Note: Follow the drawings for exact reference level.
- Epoxy-anchor starter bars into drilled holes in the plinth, aligned with the formed steps and with the first concreted step.
  Why/Where: To tie the steps to the plinth.
- Complete step reinforcement with welded-mesh panels as per the strengthening design.
- Pre-wet the hardened concrete and timber formwork ~2 hours before the next concrete pour.
- Pour and compact concrete in the remaining steps (Figure 45n).
- Strip forms after ~48 hours and cure (prevent moisture loss).

B) Render (plaster) the exterior plinth
- Bush-hammer the concrete plinth surface, then render the entire exterior plinth with water-proof mortar around the full perimeter.

C) Perimeter sidewalk
- Excavate a strip about 25 cm deep along the building perimeter, width = sidewalk width.
- Mechanically compact and level the subgrade.
- Set kerbs on a stabilized ballast bed (with cement) around the sidewalk outline, string-line aligned to the design level.
- Mark the finished sidewalk level on the rendered plinth (a control line), ensuring the slope falls toward the kerb.

- Fix inverted-U steel loops (rebar "U"s) along the perimeter, ~every 2 m, in the plane of the plinth.
  Why/Where: They hold the temporary plywood edge formwork.
- Install a foil-wrapped plywood edge formwork clipped to those steel loops, with its top edge exactly on the mark on the plinth.
  Where: This acts as the inner-screen guide for the sidewalk.
- Place a 15 cm capillary gravel layer between kerb and plinth; level it.
- Lay PVC sheet over the gravel (moisture barrier).
- Reinforce the sidewalk with welded-mesh panels on spacers; lap 3 mesh squares.
- Divide the sidewalk into bays ("plots") using plywood strips at the spacings shown in the drawings (corresponding to the strengthening project given by the engineer).
- Pour concrete in alternating bays, compact, and screed using the foil-wrapped plywood edge and the kerbs as guides.
- Wet and cover the finished surfaces to cure.
- After the first bays have completed their initial shrinkage, pour the remaining bays the same way; wet and cover to cure.
- Remove the steel loops, cut and remove the foil, and take off the plywood edge along the plinth; then pull out the remaining foil.
- Remove the plot dividers.
- Fill the joints between bays with hot bitumen.

Element map (where each thing sits, in section):

- Plinth: the lower exterior band of the building wall, just above ground level.
- Stair foundations: trenches and RC footings adjacent to the plinth; reinforcement ties into the building foundation via epoxy-anchored dowels.
- First tread/landing slab: gravel + PVC + mesh over compacted fill, tied into the foundation.
- Step risers & treads: RC built upward from the landing to the design level.
- Perimeter sidewalk: from plinth (inner edge) to kerb (outer edge) → gravel (capillary break) + PVC + mesh + concrete, sloped away from the building.
- Joints between bays: bitumen-filled after curing.

5.4.17 Execution of Exterior and Interior Plastering on Confined Masonry Walls (One-Brick Thick) around the Entrance Hall

This work is similar to plastering of TFM walls whose masonry bonds were reinforced with welded meshes.

5.4.18 Strengthening the Roof Structure

Quick glossary (what is what):

- Roof covering — The outer, weatherproof layer of the roof (tiles, metal sheets, shingles) removed down to the structural framework.

- Roof framework — The structural timber system (trusses, rafters, purlins) that supports the roof covering.
- Antiseptic/fireproof-treated timber — Wood treated against biological attack (fungi/insects) and to improve fire resistance.
- Steel strap plates & bolts — Flat metal connectors and through-bolts used to mechanically fix new timber segments to existing members via drilled holes.
- Supplementary supports — Added bearing points or sistered members (extra joists/rafters) in dry, treated timber to increase capacity where needed.
- Battens — Narrow strips fixed over rafters to carry the roof covering; removed here as part of upgrading the build-up.
- Roof sheathing (board decking) — Continuous timber boards or panels nailed to rafters, providing a solid substrate and diaphragm action.
- Anti-condensation membrane — Vapor-permeable, water-shedding underlay placed above the sheathing (cold side) to let interior moisture escape while blocking external water.
- Counter-battens — Vertical timber strips nailed over rafters (on top of the membrane) to create ventilation/drainage channels.
- Horizontal battens (tile battens) — Laths fixed perpendicular to rafters, spaced per the metal roofing profile, to support and fix the sheet panels.
- Profiled sheet metal roofing — Corrugated or trapezoidal metal panels installed as the new roof covering, anchored to the battens.

Procedure step-by-step:

- Complete removal of the roof covering from the roof framework.
- Identification of deteriorated elements.
- Marking deteriorated zones.
- Temporary propping of the respective elements.
- Cutting out deteriorated portions and replacing them with new segments of healthy, dried, antiseptic/fireproof-treated timber, fixed with steel strap plates and bolts through drilled holes.
- Supplementing supports for some members where needed with dried, treated timber.
- Removing the battens from rafters and replacing them with roof sheathing (board decking) nailed in place.
- Applying an anti-condensation membrane over the sheathing (on the cold side, under the roofing, allowing outward drying); then, over rafters, nail vertical counter-battens along the roof pitch and install horizontal battens at spacing depending on the profiled metal sheet roofing to be installed.
- Installing the profiled sheet metal roofing panels.

5.4.19 Dismantling the Old Ceiling, Reinstalling Thermal/Acoustic Insulation between Slab Beams, and Rebuilding the Ceiling with Modern Elements

Quick glossary (what is what):

- Walking planks — Temporary boards laid to create safe access/working paths at height (e.g., under a slab soffit).
- Slab soffit — The underside of a floor/roof slab (the "ceiling" face when viewed from below).
- Beam soffit — The underside face of a beam.
- Deck planks — Boards forming the underside of the floor above (or roof deck), spanning across beams/joists.
- Vapor barrier/vapor retarder (warm-side installation) — A membrane placed on the interior (heated) side of insulation to limit water-vapor flow into colder layers, reducing condensation risk.
- Mineral-wool thermal insulation — Non-combustible fibrous insulation (stone or glass wool) installed between beams to reduce heat loss and improve acoustic performance.
- Support for insulation: galvanized wire or rabbit mesh — Corrosion-resistant wires or light wire mesh (aka "chicken wire/rabbit mesh") fixed to beam sides/undersides to hold insulation in place.

Procedure step-by-step:

- Bringing walking planks for access to the slab soffit.
- Removing remaining rough-sawn timber strips (≈ 7 × 5 cm) nailed to beam soffits.
- Cleaning the beam surfaces.
- Antiseptic and fire-retardant spray treatment of the soffit surfaces (deck planks + beams).
- Vapor barrier/vapor retarder is installed on the warm (interior) side of the insulation to reduce moisture migration into colder areas.
- Installing modern mineral-wool thermal insulation between beams, supported by galvanized wire nailed to beams or by rabbit mesh.
- Rebuilding the ceiling structure with modern elements.

6 House Comfort Details

6.1 THERMAL INSULATION PROPOSALS FOR TIMBER-FRAMED HOUSES

The demand for heating and cooling has increased considerably in recent decades, as a result of the growing and changing needs for comfort, living, and working conditions, industrialization, and many other factors.

One of the factors causing temperature fluctuations in buildings is heat transfer through the building envelope by conduction and radiation. This heat loss or gain can be minimized through proper thermal insulation.

The physical, mechanical, and thermal properties of insulation materials include density, moisture content, compressive strength, tensile strength, but very important are thermal conductivity and thermal inertia.

Thermal insulations are of three types:

- **Synthetic insulation**, such as expanded polystyrene, extruded polystyrene, polyurethane foam, etc.
- **Mineral insulation**, e.g., glass wool, rock wool;
- **Natural insulation**, e.g., sheep's wool, cellulose, hemp, flax, straw (Figure 46), cork, etc.

In general, the thermal conductivity is about 0.040 W/(m·K), with small differences, for all common types of insulation materials. However, those from renewable sources, under realistic conditions, may have enhanced performance.

The advantages of thermal insulation materials from renewable sources are vapor permeability, renewable raw materials, mass and thermal inertia, recyclability, and potentially low embodied energy.

When applying thermal insulation to traditional houses, which are generally made of natural materials (timber, clay/earth, etc.) we have to consider preserving the breathability of the house, such as the natural materials to manage the moisture naturally, and avoid vapor-tight barriers. Also, we have to avoid internal condensation, and we can do this by choosing materials with good capillary activity and vapor diffusion properties.

DOI: 10.1201/9781003647249-6

FIGURE 46 Semi-rigid straw panels (left) applied on a house with diagonal planks (as in TRAROM solution) (right).

One solution, using natural materials, may be with semi-rigid straw panels. The thickness of the panels applied can be the same as for the expanded polystyrene (i.e. 8 cm), with a 2 cm plaster on top of it on each side of the house.

Besides the insulation applied on the walls, which may be exterior (preferred) or interior, depending on the project's limitations, the roof and floor also need to be insulated. The roof should have insulation between the rafters and an airtight but vapor-open membrane on the warm side. The floor should also have insulation below the timber decking and maintain capillarity break and ventilation under the floor.

6.2 FINISHES

On all types of structures, finishes can be either dry or wet.

Wet finishes, conventional ones, with pre-mixed products (plasters), specific recipes, or those prepared on-site, must meet certain characteristics. Vapor permeability is one of the basic requirements. This must be respected so that indoor water vapor can move toward the outside, thus preventing mold growth. Another requirement is that the interior finish must be at least as permeable as the exterior one.

Materials can be used for reinforcing mortars—such as galvanized wire mesh, fiberglass mesh, reed mesh—or solutions that provide plasters with increased resistance and therefore reduced maintenance.

Dry finishes can also be used, of many types, interior or exterior, made of wood, gypsum board, composite materials, etc. The vapor permeability requirement also remains valid in this case.

However, it is recommended to apply plaster mortars on the exterior, over the insulation. This closure offers protection of the wood against weathering and fire risk. This finish also provides sealing, which is essential for thermal efficiency.

Specifically for traditional timber frames with infills, the finishes must be breathable, capillary active (able to absorb and release moisture safely), compatible with timber, lime, and clay and flexible (to accommodate small movements without cracking). We should avoid modern cement renders, acrylic paints, or synthetic sealants—they trap moisture, leading to rot or frost damage in timber and infills.

Further, suitable finishes for timber frames with infills are listed:

a. Lime Plaster / Lime Render
- Traditional and ideal choice for masonry or earth infills.
- Allows moisture to evaporate, prevents cracking due to flexibility.
- Can be finished smooth or textured.
- Often applied in 2–3 coats (base, float, finish).

  It is composed of lime putty or hydraulic lime + sand (sometimes with hair or fiber reinforcement).

  It can be applied as limewash, mineral paint, or left natural.

b. Clay Plaster
- Best for wattle-and-daub or earth-based infills.
- Excellent humidity regulator; fully reversible.
- Apply on reed mat or directly onto clay substrate.
- Finish with clay paint, limewash, or natural wax coating.

  It is not suitable for exterior faces unless well protected by large roof overhangs or timber cladding.

c. Lime-Hemp or Lime-Cork Render
- Modern "eco-heritage" renders combining lime binder with hemp shiv or cork granules.
- Adds insulation and retains breathability.
- Good for irregular surfaces and gentle retrofits.

d. Traditional Earth Render with Straw/Chaff
- Historic method for clay-infill walls.
- Must be sheltered by **wide eaves** or **timber weatherboarding** to withstand rain.
- Can be limewashed for protection.

In Table 4, other possible finishes are described.

TABLE 4
Other finishes (for the exposed timber frames)

Type	Description	Moisture Behavior	Maintenance	Comments
Natural linseed oil	Traditional penetrating oil	Breathable	Re-apply every 3–5 years	Deepens color, enhances grain
Tung oil or natural oil blends	Similar to linseed, faster drying	Breathable	Periodic	More UV-resistant
Silicate or mineral stains	Chemically bond to wood, very durable	Highly vapor-open	Low	Suitable for façades
Breathable microporous wood stains (water-based)	Modern but compatible if vapor-open	Moderate	5–8 years	Widely available
Limewash on timber	Traditional in some regions	Very vapor-open	Frequent	Low cost but chalky finish

To avoid:

- Film-forming varnishes, acrylic or polyurethane paints—they seal the surface and trap moisture.
- Bituminous or tar coatings—they block diffusion and are irreversible.

Related to the traditional color finishes some solutions may be:

- Limewash: classic finish, applied in several thin coats, mildly antiseptic.
- Mineral silicate paints: modern equivalent—durable, UV-stable, fully breathable.
- Clay paints: for interiors, beautifully matte and healthy indoor climate.

6.3 GLAZED SURFACES

Another important component of a building envelope is the joinery (windows and doors). Both the frames and the glass must be as efficient as possible, with low thermal transfer. There are many types of frames: wood, aluminum, PVC, or composites, produced with increasingly better performance. Glass, made of multiple panels, is also necessary for reducing energy losses, sound insulation, and, when required, protection against radiation.

7 Conclusions

This book aims to provide the reader with technological details regarding the solutions or options that owners and engineers have when willing to strengthen their traditional timber frame with different types of infills.

Based on our experimental background we described several strengthening solutions that were tested in the laboratory and compared in terms of contribution in strength and stiffness (red line vs. black line in Tables 1 and 2). We considered both dry and wet solutions, the latter being more effective, even though more time consuming and generally more expensive.

In Chapter 5, we have presented a case study on a classical traditional house in Romania, which we personally investigated. We indicated the dimensions of the structural elements, the component materials, and the damages observed. Then we recommend strengthening and the steps taken to implement the strengthening solution proposed, so the reader can understand the methodology and approach for such works, in case they want to do it themselves (although not recommended, instead, please contact an engineer).

We insist highlighting the fact that even if we described concrete based solutions for the case study strengthening, our approach is considering the strength contribution firstly, so in case of an earthquake prone area (the case of Romania), this issue should be prioritized. But in lower seismicity areas, the natural material based solutions shown in Chapter 4 can also be adopted. Our aim was to provide the reader with correct and scientifically proven solutions, together with their correct application, which in our opinion is equally important, to reach the effectiveness of the proposed and tested strengthening solutions.

Of course, there is more work to do on the topic, and while the new materials development is faster than our research funding, we must consider further the compatibility of the strengthening solutions to the existing ones, so we can prevent unwanted impact (long term) on our houses' comfort. If we forget about capillarity action prevention, we may be comfortable for five years, but later, the mold and biological decay will become a problem which is very costly, as it requires a complex intervention, sometimes impossible.

DOI: 10.1201/9781003647249-7

References

[1] A. M. Gonçalves, J. Gomes-Ferreira, L. Guerreiro, and F. Branco, "Seismic retrofitting of timber framed walls," *Materiales de Construccion*, vol. 64, no. 316, October 2014, doi: 10.3989/mc.2014.06913.

[2] A. Charleson, *Seismic Strengthening of Earthen Houses Using Straps Cut from Used Car Tires: A Construction Guide*, September, 2011. Available: https://world-housing.net/wp-content/uploads/2011/06/Seismic_strengthening_of_earthen_houses_using_tire_straps_English2.pdf

[3] G. J. Sechi, C. Chesi, and L. Martinelli, "Numerical modeling of the seismic response of adobe houses protected with used car tyre straps," In *SAHC2014–9th International Conference on Structural Analysis of Historical Constructions*, 2014.

[4] F. F. Khorasani and M. Z. Kabir, "Experimental study on the effectiveness of short fiber reinforced clay mortars and plasters on the mechanical behavior of adobe masonry walls," *Case Studies in Construction Materials*, vol. 16, 2022, doi: 10.1016/j.cscm.2022.e00918.

[5] N. Sakthieswaran and M. Sophia, "Prosopis juliflora fibre reinforced green building plaster materials — An eco-friendly weed control technique by effective utilization," *Environmental Technology & Innovation*, vol. 20, p. 101158, 2020, doi: 10.1016/j.eti.2020.101158.

[6] F. Peña Mondragón and P. B. Lourenço, "Criterios Para El Refuerzo Antisísmico De Estructuras," *Revista de Ingeniería Sísmica*, vol. 66, no. 87, pp. 47–66, 2012, doi: 10.18867/ris.87.45.

[7] T. L. G. Michiels, "Seismic retrofitting techniques for historic adobe buildings," *International Journal of Architectural Heritage*, vol. 9, no. 8, pp. 1059–1068, 2015, doi: 10.1080/15583058.2014.924604.

[8] E. Poletti, G. Vasconcelos, and M. Jorge, "Application of near surface mounted (NSM) strengthening technique to traditional timber frame walls," *Construction and Building Materials*, vol. 76, pp. 34–50, February 2015, doi: 10.1016/j.conbuildmat.2014.11.022.

[9] D. Torrealva, J. Neumann, and M. Blondet, "Earthquake resistant design criteria and testing of adobe buildings at Pontificia Universidad Catolica del Peru," *Proceedings of the Getty Seismic Adobe project 2006,* pp. 3–10, 2006, [Online]. Available: www.getty.edu/conservation/publications_resources/pdf_publications/pdf/gsap_part1a.pdf

[10] C. M. Irimia and R. Sofronie, "Reinforcing original masonry with polymer grids," *IOP Conference Series: Earth and Environmental Science*, vol. 664, no. 1, 2021, doi: 10.1088/1755-1315/664/1/012077.

[11] A. A. Katsamakas and M. Vassiliou, "Lateral cyclic response of deformable rolling seismic isolators for low-income countries," in *3ECEES Bucharest*, 2022, doi: 10.3929/ethz-b-000570039.

[12] A. A. Katsamakas, M. Chollet, S. Eyyi, and M. F. Vassiliou, "Feasibility study on reusing tennis balls as seismic isolation bearings," *Frontiers in Built Environment*, vol. 7, September 2021, doi: 10.3389/fbuil.2021.768303.

[13] A. Tsiavos, A. Sextos, A. Stavridis, M. Dietz, L. Dihoru, and N. A. Alexander, "Experimental investigation of a highly efficient, low-cost PVC-Rollers Sandwich (PVC-RS) seismic isolation," *Structures*, vol. 33, May, pp. 1590–1602, 2021, doi: 10.1016/j.istruc.2021.05.040.

[14] A. Dutu and Y. Yamazaki, *Seismic Resistance of Vernacular Timber Frames with Infills*. Boca Raton: CRC Press, 2024, doi: 10.1201/9781003405375.

[15] A. Dutu, M. Niste, I. Craifaleanu, and M. Gingirof, "Construction techniques and detailing for Romanian paiantă houses: An engineering perspective," *Sustainability (Switzerland)*, vol. 15, no. 2, p. 1344, 2023, doi: https://doi.org/10.3390/su15021344.

[16] A. Dutu, "StrongPa Project – Appropriate strengthening methods for the Romanian traditional "paiantă" house," project number PN-III-P2-2.1-PED-2021-1428 – Experimental Demonstrative Project (UEFISCDI Romania), budget 130 000 euro, 2022–2024. Website: https://tfmro.utcb.ro/infilled-timber-frames/strongpa-activities/, accessed on March 27, 2026.

[17] A. Dutu, M. Niste, I. Spatarelu, and D. I. Dima, "TFMRO project (Seismic evaluation of Romanian traditional residential buildings)." [Online]. Available: http://tfmro.utcb.ro/

[18] A. Dutu, M. Niste, I. Spatarelu, D. I. Dima, and S. Kishiki, "Seismic evaluation of Romanian traditional buildings with timber frame and mud masonry infills by in-plane static cyclic tests," *Engineering Structures*, 2018, doi: 10.1016/j.engstruct.2018.02.062.

[19] S. Mey, I. Hirano, A. Dutu, M. Niste, D. Mocanescu, and S. Kishiki, "Extensive experimental study on five types of paianta walls – StrongPa project," *Engineering Structures*, vol. 331, May 2025, doi: 10.1016/j.engstruct.2025.120003.

[20] N. Sathiparan and K. Meguro, "Shear and flexural bending strength of masonry wall retrofitted using PP-band mesh," *Constructii*, vol. 14, no. 1, pp. 3–12, 2013, [Online]. Available: https://constructii.incd.ro/Archive/2013-1/Constructii_2013_Vol.14_No.1_ID2013140101.pdf

[21] S. Efthymiopoulos and Y. Aktas, "TG 26_2024 Mould in buildings," Published by BSRIA, ISBN: 978-0-86022-816-5, December 2024. [Online]. Available: www.bsria.com/uk/product/rMoj7D/mould_in_buildings_tg_262024_a15d25e1/?srsltid=AfmBOoonZCoiIYuj0H2z6iO5wb2_Q-Y5tKsTUDkAKDpWlVgp_FUPv4pk

Index

For Product Safety Concerns and Information please contact our EU representative GPSR@taylorandfrancis.com Taylor & Francis Verlag GmbH, Kaufingerstraße 24, 80331 München, Germany

Batch number: 10777354

Printed by Printforce, the Netherlands